Pricing of Ecosystem Services in Agriculture: A Basis of Crop Insurance

Pratap Bhattacharyya • Priyabrata Santra •
Debashis Mandal • Biswajit Mondal

Pricing of Ecosystem Services in Agriculture: A Basis of Crop Insurance

Springer

Pratap Bhattacharyya
Crop Production Division
Indian Council of Agricultural Research
(ICAR)- National Rice Research Institute
(NRRI)
Cuttack, Odisha, India

Priyabrata Santra
Natural Resources Division
Indian Council of Agricultural Research
(ICAR)-Central Arid Zone Research Institute
Jodhpur, Rajasthan, India

Debashis Mandal
Soil Science & Agronomy Division
Indian Council of Agricultural
Research (ICAR)-Indian Institute
of Soil and Water Conservation
Dehradun, India

Biswajit Mondal
Social Science Division
Indian Council of Agricultural Research
(ICAR)- National Rice Research Institute
(NRRI)
Cuttack, Odisha, India

ISBN 978-981-19-4415-4 ISBN 978-981-19-4416-1 (eBook)
https://doi.org/10.1007/978-981-19-4416-1

This Springer imprint is published by the registered company Springer Nature Singapore Pte Ltd.
The registered company address is: 152 Beach Road, #21-01/04 Gateway East, Singapore 189721, Singapore

Foreword

डॉ. हिमांशु पाठक
सचिव, एवं महानिदेशक
Dr HIMANSHU PATHAK
SECRETARY (DARE) & DIRECTOR GENERAL (ICAR)

भारत सरकार
कृषि अनुसंधान और शिक्षा विभाग एवं
भारतीय कृषि अनुसंधान परिषद
कृषि एवं किसान कल्याण मंत्रालय, कृषि भवन, नई दिल्ली 110 001
GOVERNMENT OF INDIA
DEPARTMENT OF AGRICULTURAL RESEARCH & EDUCATION (DARE)
AND
INDIAN COUNCIL OF AGRICULTURAL RESEARCH (ICAR)
MINISTRY OF AGRICULTURE AND FARMERS WELFARE
KRISHI BHAVAN, NEW DELHI 110 001
Tel.: 23382629; 23386711 Fax: 91-11-23384773
E-mail: dg.icar@nic.in

According to the Millennium Ecosystem Assessment, every 15 out of 24 (approximately 60%) of the global ecosystem services are being degraded. Degradation of these services will be much faster in the future as we are exceeding many planetary boundaries such as climate change, biodiversity loss and biogeochemical flows of elements. Nearly 99% of human food comes from land surface and soil. Soil and water are coming under rising pressure because of misuse and mismanagement. The accelerating use of natural resources is continuing to affect key ecosystem services (ES), threatening sustainability and food security.

It is required to understand the various services that ecosystems provide in support of human wellbeing. Massive efforts are being made at global level for developing framework and guidelines for sustaining natural resources in support of rural transformation and inclusive growth through participation of primary stakeholders. A supportive step in this direction will be the formulation of a structured approach and methodology that was lacking to compute values of different ecosystem services benefited through adoption of soil and water conservation measures and renewable energy management. This book entitled *Pricing of Ecosystem Services in Agriculture: A Basis of Crop Insurance* is a rich and important reference source for students, researchers and policymakers in sustainable land management, ecology, environmental studies, ecological economics and sustainable development. The cross-cutting, thematic chapters challenge conventional wisdom in some areas and validate new methods and approaches for sustainable land and energy management. I congratulate the authors for their painstaking efforts in bringing out this comprehensive document which is most relevant in the present context. I hope that this publication will contribute to sustainable financing for conservation and restoration of ecosystems around the world.

New Delhi, India

Himanshu Pathak

Preface

Soil management and agricultural practices provide many ecosystem services that are of value to the public welfare, but for which there are few or no developed markets. Therefore, we thought that there is a need for a book *Pricing of Ecosystem Services in Agriculture* that should be easy to understand and written in a lucid manner for students, researchers and policy makers. We strongly believe that the valuation of ecosystem services in agriculture must be linked with 'Crop Insurance'. The book presents the concept and methodologies of pricing of ecosystem services in agriculture with examples and case studies. In nutshell, the book is about quantifying the ecosystem services (ESs) provided by agriculture including provisioning, regulating and cultural services and subsequently linked those to framing the crop insurances. Researchers, post-graduate students and policy makers are the targeted audience of the book. The main topic of the book revolving around the approaches of ES in agriculture: pricing of agricultural products, soil and water conservations, watershed managements, carbon sequestration, greenhouse gases emissions-mitigation, renewable-energy contribution. Subsequently, we linked the quantified ES in agriculture as a basis of crop insurances along with the critical issues of fixing premium and regulation of crop insurances. Valuation and pricing of ES in agricultural services both tangible and intangible is a challenge. Through this book we have tried to put clearcut methodology for pricing of ES in agriculture, which is highly debatable particularly in environmental and cultural services. Here, our focus is to use the appropriate methodology and to provide an indication of the types of methods that would be appropriate for deriving new-market benefit values of different ESs. We try to address the problem specific to pricing of ES in agriculture, logical and acceptable framework for crop insurance in the light of both economic and societal perspective for sustainable eco-friendly agriculture in future.

The six chapters of the book systematically covers the concept and approaches of ecosystem services in agriculture, pricing of agricultural products, soil, water conservation and management, carbon sequestration and environmental regulations. It also covers the valuation of renewable energy-based applications in agriculture, and finally linked the pricing of ES with crop insurance.

Chapter 1 begins with the concept and definition of ecosystem services (ES) as a natural capital. The importance of ES in general and agriculture in particular is described with simple diagrams. Then general methodologies of pricing of ESs in agriculture are explained. The value of soil and water resources on ecosystem service delivery is also elaborated with examples. The payment needs of ESs in agriculture are placed in a logical manner for the reader in the respective places in this section. The major challenges of pricing/valuation of ESs are pointed out as experienced and reported by several researchers and policy makers. A clearcut future need and key messages are also embedded in this chapter.

The second chapter deals with pricing of agricultural products like food, fibre and raw materials for industries. The second section of this chapter elaborates the valuation of soil nutrient management in agriculture, special reference the amelioration of problem soil (saline soil, sodic soil, acid soil). Valuation of water management in agriculture including 'water supply management system', 'agriculture water control structure', flood control and rainwater harvesting structures are briefed. The options of markets for ecosystem services in agriculture are presented in future road map of this chapter. In Chapter 3, the pricing of soil erosion control and soil build up are exhaustively discussed with examples. Almost all the components of ESs (provisioning, regulating, supporting and cultural) are presented separately in light of soil and water conservation measures with definite methodologies for their pricing. Pricing the ecosystem benefits of soil conservation measures is emphasized specifically. Valuation of 'Participatory Watershed Management' with case studies in Himalayan region of India is a valuable section of this chapter. Basic differences of valuation and monetary pricing and important economic terminologies are presented as a message at the end of this chapter.

In Chapter 4, pricing of carbon sequestration and gas regulation are elucidated. Quantification and valuation of carbon balances in rice paddies, wetlands and agroforestry are presented with elaborate methodologies and examples. Valuation of gas regulation is an important section of this chapter which describes how to quantify and value the GHGs mitigation in agriculture. Two separate sections of this chapter describe the pricing of bioremediation, pollution control and waste-water utilization in agriculture. Another elaborate section of this chapter is the valuation of residue burning and pricing of the crop-residue management options in agriculture which are presented with examples.

In Chapter 5, ESs of renewable energy applications in agriculture are discussed including solar thermal devices, solar photovoltaic (PV) systems, wind energy technologies and biomass energy. A section of this chapter is devoted to the increasing importance of renewable energy applications in the world as well as in India with the background concerns over fast-depleting fossil fuels and associated carbon footprints. Further, the methodologies to calculate ESs of renewable energy applications are discussed in detail. Case studies on pricing of ESs of few solar thermal devices, solar PV pumping system, agri-voltaic system, wind turbines and biomass energy generation are discussed. At the end of the chapter, the future roadmap for strengthening the pricing methodologies of ESs and it's adoption for providing incentives to the stakeholders are discussed.

Chapter 6 deals with basis of crop insurance, different approaches of insurances followed by different countries including India. The challenges and bottlenecks of implementing crop insurance in policy and local level are described in detail. Then linking of crop insurance with payment of ecosystem services (PES) and how crop insurance can be used as the basis of PES are explained.

We received ungrudging help from ICAR-National Fellow Project, NICRA, ICAR-NRRI, Cuttack, ICAR-IISWC, Dehradun, ICAR-CAZRI, Jodhpur, and from a number of scientists and well-wishers for preparing this book. We wish to mention Dr T Mohapatra, Dr S C Datta, Dr H Pathak, Dr P Swain, Dr M J Baig, Dr D Maity; Dr Nishita Giri, Dr Sushma Tamta, Mr P K Dash; Mr S R Padhy; Dr S Neogi; Mr Anubhab Das; Mr Anil Mistri, Dr (Mrs) S Pattanaik, Dr Mahesh Kumar, Dr P C Pande, Dr N M Nahar, Dr S Poonia, and Dr S H Majumdar. We have duly acknowledged the sources of the diagrams and tables that have been reproduced from other sources and publications.

Cuttack, India — Pratap Bhattacharyya
Jodhpur, India — Priyabrata Santra
Dehradun, India — Debashis Mandal
Cuttack, India — Biswajit Mondal

Contents

About the Authors

Pratap Bhattacharyya is currently working as ICAR-National Fellow and Principal Scientist at ICAR-NRRI, Cuttack, Odisha, India. He has done his MSc and PhD from IARI, New Delhi, India. His fields of specialization are climate change, carbon dynamics, GHGs emissions-mitigation, resource conservation technologies and microbial diversity in rice. He is Fellow of National Academy of Agricultural Sciences, India (FNAAS); Fellow of IASWC, ARRW and WAST. He received ICAR-LBS Young Scientist Award; Dr K J Tejwani Award; Mosaic Foundation Award; Certificate of Excellence on Eddy Covariance from UGA, USA. He has published more than 150 research papers, 6 books, 20 book chapters and popular articles. He has guided 5 PhD and 8 MSc students.

Priyabrata Santra is currently working as Principal Scientist in ICAR-CAZRI, India. He obtained his Masters in Agricultural Physics from IARI, New Delhi, and PhD in Soil Physics/Hydrology from IIT, Kharagpur. He visited International Centre on Theoretical Physics (ICTP) and was awarded with Junior Associate (Soil Physics) of ICTP. Dr Santra worked as visiting scientists at International Soil Research and Information Centre (ISRIC) at Wageningen, Netherland. He was awarded with FAST Track research grant from DST, Govt. of India. He received ICAR-LBS Young Scientist Award. He is working on natural resources management, digital soil mapping and use of renewable energy on agriculture. He published 64 research papers in reputed international/national journals, three edited books, two authored books and two research bulletins.

Debashis Mandal is currently working as ICAR-National Fellow at ICAR-IISWC, Dehradun. He received the PhD degree in soil science and agricultural chemistry from IARI, New Delhi. His research interests comprise assessment and monitoring of soil erosion, soil sustainability, land degradation and reclamations processes and land degradation-induced losses of productivity and ecosystem services. Dr Mandal worked as a visiting scientist at CMASC, at Ohio State University, Columbus, USA, and conducted research with Prof. Rattan Lal. He published more than 70 peer-reviewed articles in scientific journals and organized several national and international conferences on soil and water conservation.

Biswajit Mondal is working as Principal Scientist at ICAR-National Rice Research Institute (NRRI), India. He obtained his PhD in Agricultural Economics from Indian Agricultural Research Institute, New Delhi. He visited Fort Collins, USA, as an expert in the workshop on 'Threat and Benefit Assessment Methodologies' for reactive nitrogen as a part of 'Towards International Nitrogen Management System (INMS)' programme. He is working as Associate Editor of Journal Oryza. His areas of expertise are resource economics, impact analysis and on-farm evaluation of agricultural technologies. He has more than 60 national and international research publications and handled more than 20 research projects.

Abbreviations

AF:	Annuity factor
AI:	Artificial intelligence
ALCC:	Annualized life cycle cost
ANPV:	Average net present value
AVS:	Agri-voltaic system
AWCS:	Agriculture water control structure
BD:	Bulk density
CA:	Conservation agriculture
CC:	Cost of carbon
CCEs:	Crop cutting experiments
CCIS:	Comprehensive Crop Insurance Scheme
CDM:	Clean Development Mechanism
CER:	Certified emission reduction
CF:	Carbon footprint
CH_4:	Methane
CO_2:	Carbon dioxide
CVM:	Contingent valuation method
DCEs:	Discrete choice experiments
DCF:	Discounted cash flow
DOC:	Dissolved organic carbon
EC:	Electrical conductivity
EF:	Emission factor
ESP:	Exchangeable sodium percentage
ESs:	Ecosystem services
ETC:	Evacuated tube collectors
ETL:	Economic threshold level
FYM:	Farm Yard Manure
GHGs:	Greenhouse gases
GHI:	Global horizontal irradiance
GIC:	General Insurance Corporation
GR:	Gypsum requirement
GWP:	Global warming potential
IMD:	Indian Meteorological Department

IoT: Internet of things
IPCC: Intergovernmental Panel on Climate Change
IRR: Internal rate of return
KUSUM: Kisan Urja Suraksha evam Utthan Mahabhiyan
LCA: Life cycle analysis
LCB: Life cycle benefit
LCC: Life cycle cost
LUC: Land use change
MEA: Millennium Ecosystem Assessment
MNAIS: Modified National Agricultural Insurance Scheme
MOP: Muriate of potash
MSP: Minimum support price
N_2O: Nitrous oxide
NAIS: National Agricultural Insurance Scheme
NCIP: National Crop Insurance Programme
NEHR: North-eastern Himalayan region
NEP: Net ecosystem production
NH_3: Ammonia
NPV: Net primary value
NWH: North-western Himalayas
OM: Organic matter
PCIS: Pilot Crop Insurance Scheme
PES Payment of ecosystem services
PMFBY: Pradhan Mantri Fasal Bima Yojana
PV: Photovoltaic
RHP: Rainwater harvesting ponds
RS: Regulating services
RUA: Reference unit area
RWS: Reference weather station
SAR: Sodium absorption ratio
SLMP: Sustainable land management practices
SOC: Soil organic carbon
SS: Supporting services
SSP: Single superphosphate
SWC: Soil and water conservation
SWCM: Soil and water conservation measures
TEV: Total economic value
TGA: Total geographical area
TWh: Terra watt-hour
UAV: Unmanned aerial vehicles
USDA: United State Department of Agriculture
USLE: Universal soil loss equation
VAWT: Vertical axis wind turbine
VOC: Volatile organic compounds

WBCIS:	Weather-Based Crop Insurance Scheme
WSMS	Water supply management system
WTA:	Willingness to accept
WTP:	Willingness to pay

1 Concept and Approaches of Ecosystem Services in Agriculture

1.1 About Ecosystem Services

The term 'ecosystem services' first coined and published by Ehrlich and Mooney (1983) (Costanza et al. 2011; Kubiszewski et al. 2020). 'Ecosystem services' (ESs) refer the benefits that human being is derived from the effective functioning of the ecosystem. In other words, ecosystem functioning and/or processes those directly or indirectly contribute to human well-being are called ESs. However, all ecosystem functions and processes are not 'ecosystem services' (ESs). Only those functions contribute to human well-being and could not be defined independently qualify as ESs (Costanza et al. 1997b, 2011; Cabral et al. 2016). The key fundamental services like survival, health, protection, and livelihoods are provided by ecosystem (Costanza et al. 1997b; Millennium Ecosystem Assessment (MEA) 2005; Sukhdev et al. 2010).

Very often these services are described or discussed in qualitative term putting more emphasis on societal well-being in non-quantitative and non-economic term. We are putting them in an economic term as 'natural capital' (Costanza et al. 2011, 2014) having a flow of services overtime. Costanza and Daly (1992) first used the term 'natural capital' for valuing the ESs (Costanza et al. 2014). The 'natural capital' is derived from those ESs which are not built or maintained by human activities, i.e. natural ecosystem and its products. However, for getting the actual benefit of 'natural capital' it must be combined with other forms of capital ('built', 'human' and 'social' capital) which require human intervention to build and maintain the stock. Therefore, ESs are the relative contribution of 'natural capital' individually or in symphony to other three forms of capital for human well-being (Fig. 1.1).

Based on the nature of services, the ESs are classified into four broad categories (Millennium Ecosystem Assessment (MEA) 2005; Costanza et al. 2014). The four categories are briefed below and the details of each of them will be discussed in the subsequent chapters of this book.

P. Bhattacharyya et al., *Pricing of Ecosystem Services in Agriculture: A Basis of Crop Insurance*, https://doi.org/10.1007/978-981-19-4416-1_1

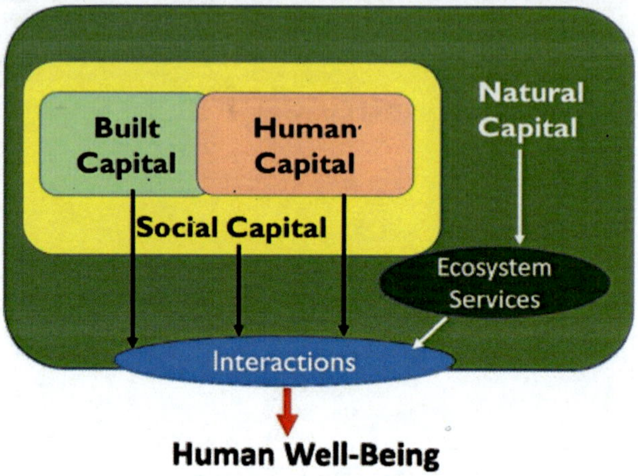

Fig. 1.1 'Natural Capital' derived from ecosystem services and its interactions with the other capitals contributing to human 'Well-Being'

- *Provisioning services*: Ecosystem services that provide food, fodder, timber and raw material for any production system and industries or 'other provisioning' benefits. Those are usually made up of 'human', 'built' and 'social' capitals. For example, rice consumed by peoples as food requires rice-farmer (human capital), rice mill (built capital), and farming community (social capital) along with natural resources like land (soil), water, oxygen, carbon dioxide, etc.
- *Regulating services*: Ecosystem services which regulate different aspects of natural and human resources/processes on integrated systems. These services have higher values in society, however, not generally marketed. These services include water regulation and purification, flood control, human disease prevention and protection, cyclone protection, air quality maintenance by indulging all the three forms of capitals with 'natural capital'. For example, the cyclone-protection services to vulnerable coastal belts require weather forecasting and cyclone-protection shelter (built capital); evacuating force (to shift the vulnerable people to safe place; (human capital)); social awareness (social capital); and natural mangrove forest cover in coastal wetland (acts as natural protector of cyclone; (natural capital)). So, here the natural mangroves cover provides the huge regulation services towards protection of cyclone.
- *Supporting services*: The ecosystem services which support and maintain the basic ecosystem functions and interlinked processes like nitrogen and carbon fixation, formation of soil and maintenance of soil health, sustaining the habitat of flora and fauna. These services influence the human well-being by supporting other related regulating, provisioning and cultural services directly or indirectly. Often these services are not combined with built, human and social capital economically but passively contribute to human benefit. For example, mitigation of greenhouse gases emission from agriculture is a 'climate regulation' service which is derived from higher primary production from agroforestry (provisioning service) combined with human (plantation and maintenance work), built (establishment of well-designed agroforestry system) and social (public awareness and

participation) capitals. However, 'climate regulation' sometimes also refers as 'ecosystem functions'.

- *Cultural services:* The services which enhance quality of life of human being are termed as cultural services. These services primarily include aesthetic, recreation, cultural identity, etc. All the three forms of capitals (built, human, social) are generally combined to provide these services. For example, in agricultural sector, a recreational service (grassland and orchards) needs a natural asset (a landscape), along with a built-up facility (land shaping with road, sodding and planting), human capital (people to maintain the grassland and orchard and appreciate the beauty) and social capital (eco-friendly nature lover tourists).

1.2 Importance of Ecosystem Services in Agriculture

Agriculture provides basic needs of human being in terms of food, fibre and raw materials for industries. These are both tangible and non-tangible benefits to ecosystem from agricultural sector that needs to be strengthened and enhanced. Several benefits for human being obtained from agriculture as ecosystem services, are both marketable and non-marketable. In general, provisioning services like food, fuel, fibre and by-products; regulatory services like biocontrol of pest, soil erosion moderation, nitrogen fixation, pollination, sustaining carbon flow, soil fertility maintenance, nutrient cycling and hydrological flow; and cultural services like cultural and recreational benefits are obtained from agriculture (Millennium Ecosystem Assessment (MEA) 2003, 2005; De Groot et al. 2012) (Fig. 1.2). However, quantity and quality of ESs from agriculture primarily depend on the inputs used for

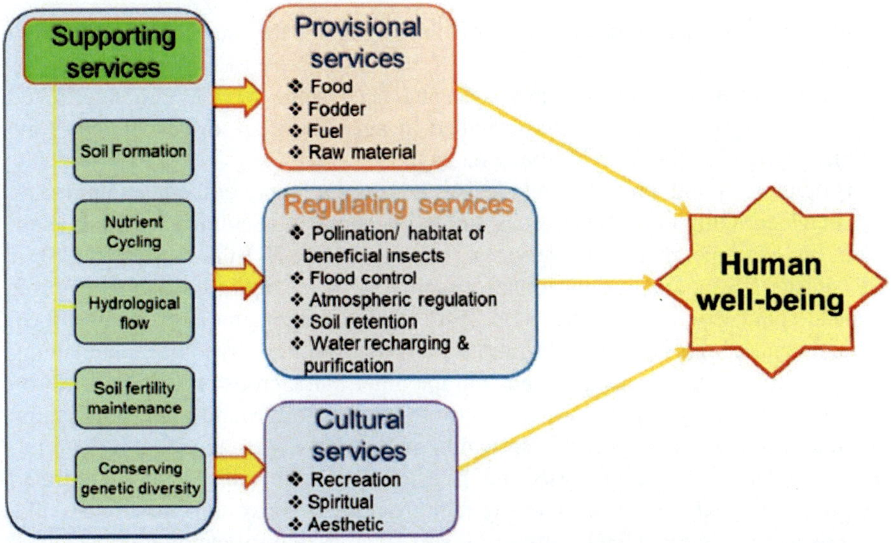

Fig. 1.2 Interlinking of different ecosystem services in agriculture (Sources: MEA 2003; Zhang et al. 2007)

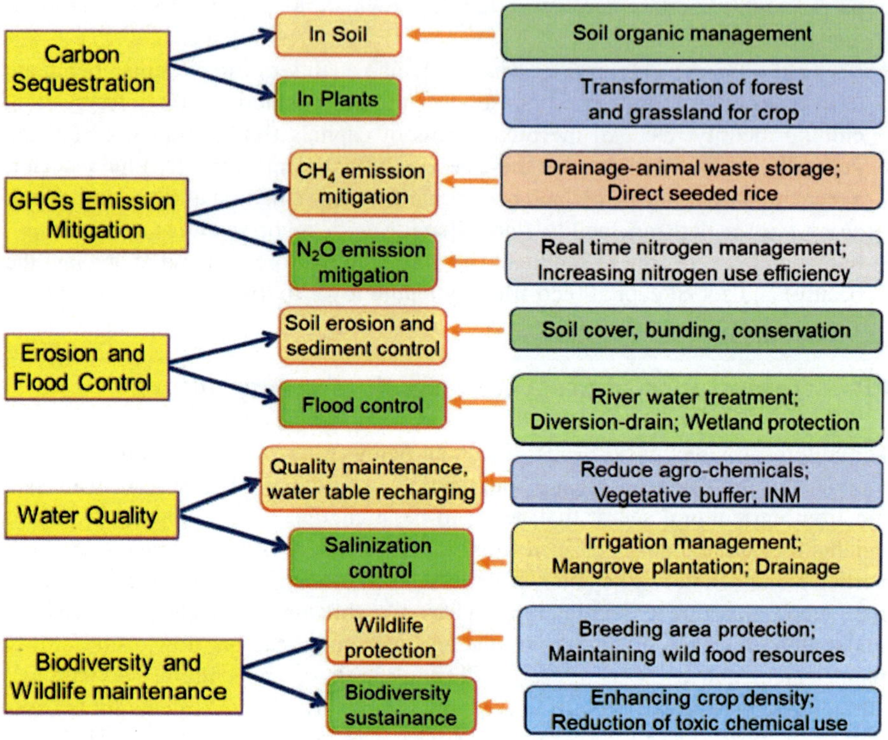

Fig. 1.3 Interlinkage of farm management practices and ecosystem services in agriculture

production and management practices followed by the stakeholders (largely farmers).

Broadly, farm management practices and ecosystem services are interlinked (Fig. 1.3) and the extents of ESs provided in agriculture are driven directly how optimally/precisely the farm has been managed.

Agriculture is both provider and receiver of ESs and many of them are unnoticed and unvalued. Only those services are limited become apparent and we think about those. For example, pollination services, (i.e. worth US $75 Billion in 2007, USDA 2007 in the USA); keeping beneficial insects (e.g. *Coccinellid* beetles predator of soybean-aphid; could lower the yield up to 25%; Costamagna and Landis 2006); wetlands and streams that reduce nitrate pollution (reducer hypoxia), maintaining water quality by shrimp-fisheries; etc. On the other hand, crops on field depends on services provided by neighbouring ecosystem in terms of groundwater, pollination, soil nutrients, gas exchanges, etc. Basically, modern concept is to recognize agriculture on a landscape scale, based on holistic enterprise rather than field-based activities only. At the same time, agriculture-related ecosystem disservices like groundwater depletion, GHGs emissions, soil erosion, desertification, marine eutrophication, soil fertility depletion, etc. are also should be addressed adequately through technological and policy interventions. In this perspective, conservation agriculture, restoration of abandon lands (including wetlands), groundwater recharging, etc. can play pivotal role for sustaining ecosystem balance.

The details of ESs provided by agriculture in terms of provisional (food, fodder, fibre, fuel, raw materials for industries, by-products), regulatory (gas and water regulation, erosion control, pollination, flood control, etc.), supporting (soil fertility, soil formation, nutrient cycling, hydrological flow, etc.) and cultural (aesthetic, recreation, etc.) will be discussed in subsequent chapters of this book with accounting and pricing along with the challenges for their economic valuation.

1.3 Value of Soil and Water Resources and Ecosystem Service Delivery

Soils play a key role in providing the ecosystem services. In the middle of 1990s, the concept of valuation of ESs started with less focus on soils' contribution towards ecosystem sustainability. The ESs provided by soils, primarily depend on soil processes and their interactions with surroundings and 'soil-use-management practices'. Soil erosion, soil loss below a tolerant limit, and compaction reduce the soil microbial diversities and carbon status which lead to land degradation that has serious consequences on ecosystem sustainability and food security. Soils as natural resources are considered as a kind of natural capital which have significant contribution for providing various ecosystem services. Around 90–99% of animal and human food sources come from the soil-plant systems (Pimentel 2006). Soils help in cleaning, filtering and regulating the drinking water; they provide essential nutrients to plants and microbes (Daily et al. 1997); they are the reservoir of organic matter and biota. Soils store three and five times of carbon compared to biosphere and atmosphere, respectively (Science 2004; Bellamy et al. 2005; Scharlemann et al. 2014). They play important role in flood control by regulating the terrestrial water flows. Therefore, maintaining the soil processes is necessary to sustain the functioning of the soils that provides the vital ecosystem services. However, soils are under tremendous pressure from human-induced activities and climate change consequences. One of the culprits of soil degradation is the exclusion of the pricing/valuation of ESs provided by soils in policy framework and land management decisions. To address this issue, we propose a valuation framework for soils-ESs and then illustrate how the soil and its various functions that provide ES can be priced in this chapter (Chap. 1) and also in the Chap. 3 of this book.

A multidisciplinary approach is needed to value the ESs contributed by soils taken into consideration the economic, societal and planetary aspects of sustainability. The valuation methods should be easy to understand and written in lucid language. As discussed earlier, the valuation method should include both the economic benefits and environmental gains, i.e. well-being of human as a whole and a population or society and in public health in specific. The valuation of soil natural capital is important to convey the need of long-term soil protection to policymakers/land managers, and farmers. The beneficiaries of ESs are varied in local and regional scale. The relationship among the independent ES at different scale is complex. Often the providers of soil-ESs are not the direct beneficiaries of those services (Breure et al. 2012; Rutgers et al. 2012). For example, services provided by soils including natural disease suppression, water supply and nutrient retention are valued predominantly, while services like gas regulation, groundwater recharging, water purification, and climate regulation are rarely priced. On the other hand, natural decomposition of organic matter and soil carbon transformation are rarely weighed by farmers, but the regional and national policy level carbon sequestration or carbon credit is the most important ones. Contrary, the soil aggregation leads to improvement of soil structure is seemed to be important for farmers, where the national policy framework officers give a little value to the soil physical structure compared to soil chemical fertility. Therefore, it is clear that the valuation of ESs provided by soil by the different stakeholders depended on scale, and the awareness of the beneficiaries about the related soil functions. So, evaluation of specific soil functions must be done in terms of economical, sociological, legal and environmental perspective. Technical knowledge is essential along with real-time information and integrated approaches for optimal soil use. Prior knowledge of the decision maker is important. Apart from that prioritizing the alternatives, land-use pattern, consideration of climate change issues and levelling of offsets are necessary for effective soil management practices for providing sustainable ecosystem services. A schematic diagram of the relationship among different soil functions, ecosystem services, policy framework and decision-making through model-based toolboxes is shown in Fig. 1.4. Understanding the dynamic nature of soil processes and enhancing the soil natural capital are the key for sustaining ecosystem services.

1.3.1 The Dilemma

Restoration of biodiversity and ecosystem services could have synergistic or conflicting aims. Land-use decision and management practices play the crucial role for getting the synergism (Power 2010; Bullock et al. 2011). For example, in grassland ecosystems, it was reported that better management could promote biodiversity as well as provide regulating and provisioning ESs. Restoration of long-term biodiversity was achieved by cessation of chemical fertilizer use and promoting seeding of nitrogen-fixing plan-species. Those practices also helped in carbon sequestration and higher economic yield (De Deyn et al. 2011). Additionally, eco-friendly restoration could also deliver higher ecosystem benefits, like climate change mitigation, soil nitrogen fixation, improved soil aggregates, higher organic carbon storage, greater water holding in soil, etc. (De Deyn et al. 2011). On the other hand, intensive chemical fertilizer uses in grasslands facilitate the growth of specific grass species, lead to lower biodiversity. Further, plant community richness directly affects soil respiration, soil microbial diversities and organic carbon decomposition (De Deyn et al. 2008). For example, fast growing grass species stimulate the rhizospheric activities that could prime the passive carbon pools residing in soil and causes faster losses of carbon than gains. However, restorative grassland

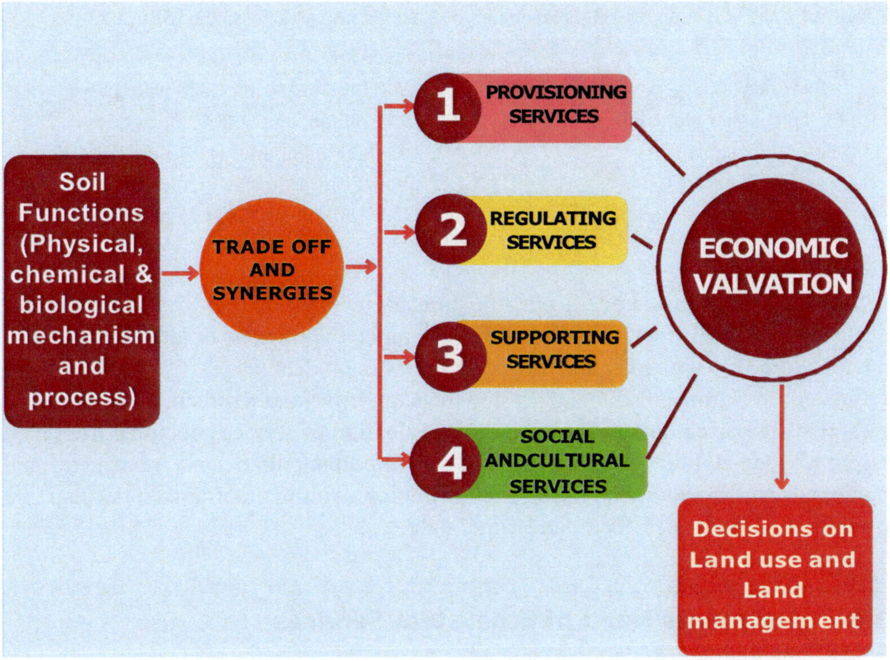

Fig. 1.4 The connection between different soil functions and ecosystem services

management practices improve the soil carbon and nitrogen contents. Moreover, those practices help in reduction of ecosystem carbon losses by respiration and the overall balance of biodiversity and ESs are positive (De Deyn et al. 2011). In this juncture, the other grassland management strategies include seeding of grass species having deep root systems, higher nitrogen and water-use-efficiencies (Lal 2004; Lal et al. 2007). There are contrasting impacts of natural capital, farming practices and soil management practices on the provision of soil services and their values. For example, the value of the ecosystem services decreased with increasing stocking rates. The decrease is higher with decreasing quality of the land. The same framework can be used to quantify the impact of soil conservation practices on the flow of ecosystem services from eroding hill pastures.

1.3.2 Valuation of Soil Ecosystem Services

Soil is an important natural capital that regulates the economic status of nations (Dominati et al. 2010). Soil natural capital is defined by the 'capacity of soil to provide the ESs required for a determined land use, assuming that sustainable practices are being used' (Hewitt et al. 2015). However, often the soil contributions are overlooked as a key resource, even though soil functions are vital to ESs (Robinson et al. 2013, 2014). Despite the huge contribution of soils towards ESs,

Table 1.1 The valuation of soil threats based on detailed impact assessment (SEC 2006)

Soil threat	Estimated annual cost
(i) Soil erosion	US $ 1.05–21.03 billion, on the 2013 price basis
(ii) Organic matter reduction	US $ 5.11–8.41 billion, on the 2013 price basis
(iii) Soil salinization	US $ 0.237–0.482 billion, on the 2013 price basis
(iv) Landslides hazards	US $ 1.8 billion, 2013 per event, on the 2013 price basis
(v) Contamination and soil pollution	US $ 3.61–25.99 billion, on the 2013 price basis

it has not been considered in classification as 'ecosystem types' for economical evaluation in TEEB (de Groot 2010). For example, the impacts of land degradations on total productivity loss and off-site environmental damages (Table 1.1) were estimated (in European soils) neglecting the supportive and regulatory ESs provided by soils that was an underestimation of the valuation of services obtained from land resources (de Groot 2010). Thus, it is important that land managers and policymakers should include soil's contribution within the frameworks that are used for evaluation of ESs in landscape scale.

1.4 Payments Need of Ecosystem Services

Pricing or valuation of ecosystem services in agriculture has recently become a critical issue both in research and policy level. It is gaining importance not only for sustaining and enhancing ES but at the same time nurturing and protecting the farmers' interest and providing a valid platform to compensating them who are maintaining/balancing these services. However, few disservices originating during practising of agriculture like soil loss, pest and disease hosting, water and fertility depletion and greenhouse gases emissions are also need to be accounted during pricing of ESs (MEA 2005; Zhang et al. 2010; Stallman 2011).

In the broader perspective, valuation of ESs is not simply to put some price to some services but to promote awareness among all the stakeholders (farmers, policymakers, civil citizens, industries, government staff and officials) regarding the concept and importance of ESs and mainstreaming those within the government policy framework. Moreover, agriculture including pastures covers 24 to 38% at earth's land area (MEA 2005), managed by human being. It is also recognized that agriculture has the good potential to provide diverse ESs in coming decades (Yu et al. 2011). Agriculture supplies all the three categories of ESs, namely provisioning, regulatory and cultural and at the same time needs supportive ESs to become productive. There is a need to pay ecosystem services provided by agriculture on educational, artistic, aesthetic and cultural including gas regulation (O_2-CO_2 balance) and maintenance of biodiversity. These are often overlooked and taken to be granted. However, unless we put values to those, either in terms of money or other incentives, the resources and approaches that provide these services will continue to be exploited. Logical pricing of ESs in agriculture would sensitize policymakers and the other stakeholders regarding the values accrued to the society as a whole and needs for maintaining those. These would promote agricultural development along with ecosystem sustainability.

Further, farmers produce several non-commodity ESs for which market does not exist, like gas regulation, groundwater recharging, soil-fertility maintenance through organic farming, etc. However, these services are valued by the society, but because of not valid system to give 'price-tag' to them, farmers are deprived, and society is also suffered for long-term perspective. This leads to a poor allocation of fund to produce those service and thus farmers under-produce the 'non-commodity' ecosystem services. These vicious cycles slowly destroy the balance of environment. Therefore, services like climate regulation, water supply, recreation, waste-treatment, and biodiversity maintenance should be valued, in order to establish economic markets for those.

In this connection market-based approaches are found more effective as it allows producer to effectively use their own information to value the non-tangible services (Freeman and Kolstad 2007). Few examples are 'sulphur-dioxide-allowance-trading-programme' (Milder et al. 2010), 'market-based environment stewardship' by USDA in the USA (USDA 2006), 'emission trading', 'eco-labelling' and 'mitigation-banking', which got positive responses and results on pricing the ESs in agriculture.

1.4.1 Specific Needs of Payments of ESs in Agriculture

- Allocation of resources to agriculture based on the present and future needs of human well-being.
- Proper financial support to landowners for supplying the ESs.
- Provide incentives to farmers to keep a portion of land in natural condition/presence of vegetation cover throughout year and/or keeping the wetland intact.
- Motivating farmers for maintaining ESs by providing optimum payment to 'non-tangible' services (regulation, supporting and cultural services) in a well-defined manner.
- Attracting policymakers towards financing ES provider.
- Create transparent and comparable markets for ESs in agriculture at local, regional and global scale.
- Above all, increase the awareness of people in general on importance of ESs in agriculture and needs for their sustaining production for well-being of human.

1.5 Challenges of Pricing/Valuation of Ecosystem Services

One of the key challenges of pricing ESs in agriculture is that it has both direct and indirect effects. So, if we put them only as on ecosystem function, then we may lead to faulty estimation. For example, soil fertility and nitrogen fixation have their direct effects on crop yield (provisioning services) and at the same time they have indirect role on soil fertility maintenance and GHGs emission mitigation (regulating services). So, at the time of pricing of the production, regulatory service also should be accounted for (Barbier 2011; Nayak et al. 2019). The major challenges of pricing of ESs provided by agriculture is defining the ownership of services and subsequent enforcement. Moreover, due to the absence of market for providing most of ESs, the farmers are getting zero price. Even sometimes consumers willing to pay but absence of market-price (price of their true ecological values), the trend of 'zero price' to many ESs continues. So, there are fewer resources allocation towards their (ESs) production at optimum level (to maintain ecological balance).

The major challenges faced in agriculture are the market opportunities to the ESs. The major reasons for this can be listed as:

- Uncertainty about environmental performance of agricultural practices like zero tillage, nutrient management, water conservation, soil protection, etc. as their performances vary with field heterogeneity and weather aberration and with the site-specific variation of implementation of package of practices.
- Uncertainty of reliable offsets to be credited in case of emission trading and GHGs mitigation markets. Non-point benefit and precise measurement are the other obstacles.
- Negotiating the pricing of long-term emission processes among different stakeholder is an issue. For example, it is difficult to decide a price for the long-term economic benefit of investing to a 'wetland-mitigation-bank'.
- Lacking proper methodologies and technical guide for fixing the value/price/credits for ESs (USDA 2007).
- Inadequate quality assurance of services; therefore, difficult to bring together the buyers and sellers and lack of coordination of conservation programme with the markets are the issues for poor markets of ESs in agriculture.

1.5.1 Double Counting of ESs in Agriculture

Double counting is a challenge in pricing of ESs in agriculture. This problem arises uncertainty and poor reliability of valuation system of ESs in agriculture. 'Double counting' means a faulty practice to count the value of an ecosystem service more than one time. Often it is happened in nation's goods and services having multiple beneficiaries at different scales. Regulatory services sometimes valued more than once in agriculture. For example, portable water supply is a provision service, while quality maintenance of water through infiltration by soil is a supporting service and surface water flow is a regulating service. Now, if we aggregate all the three services of water flow then it could lead to double counting as surface-water-flow and water quality maintenance by soil ultimately lead to final products of 'portable water'. Therefore, end product must be considered for valuation. End products are the resultant of various services of intermediate ecosystem function/processes. There are four primary methods by which double counting can be reduced. Those are:

- Choosing the final end products/benefits of ESs.
- Proper identification of spatio-temporal scale of beneficiaries of ESs.
- Development and maintaining of consistent classification of ESs.
- Selecting of site-specific, precise, and transparent pricing methodologies of ESs.

However, due to complexity of variation of ESs, it is difficult to remove double counting completely from pricing/valuation of ESs. So, clear-cut approaches to differentiate the ecosystem processes, ecosystem functions and ecosystem services would help to minimize the risk of double counting.

1.5.2 Disservices in Agriculture

Major regulating services (soil fertility, pollination, etc.) are the inputs of production of food, fodder, fuel and other raw material (provisioning services) in agriculture. Agriculture also experienced the 'ecosystem disservice' which generally increases the production cost and/or reduces productivity. For example, weed competition with desired crops for nutrients and water. These disservices depend on farm management practices and their relationship with effective utilization of other natural resources. Specifically, ecosystem disservices can be classified into two broad categories, one is the losses and damages of natural resources and other competition for resources to produce food, fodder, fibre and fuel (Fig. 1.5).

There is feedback effect of regulating services and disservices. For example, losses of beneficial insects can cause outbreaks of certain pest and disease. Similarly, nutrients losses through run-off could enhance the competition for the nutrients between crops and weeds and consequently resulted in a reduction of food production.

Loss or damage of the crop due to pest outbreak is significant. Use of synthetic chemical is often the only option to check the outbreak of pest and disease. Excessive use of pesticides could damage the beneficial insects responsible for

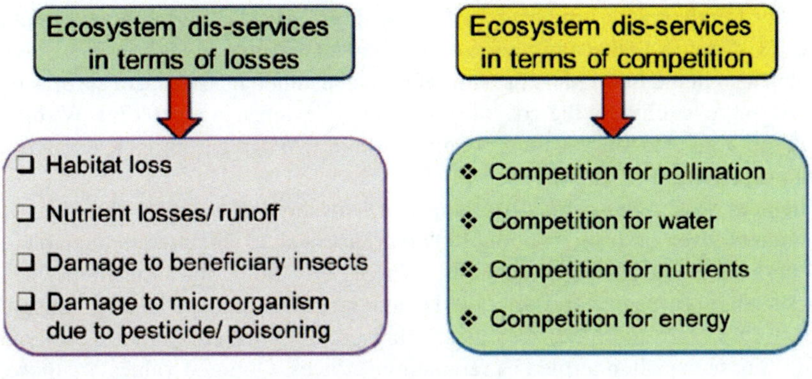

Fig. 1.5 Ecosystems disservices in agriculture

pollination or even predators of some pests. So, it further damages the ecological balance and biodiversity of the system and at the same time increases the cost of production of food and fibre. Therefore, we can see the disservices can negatively affect the other regulatory and supporting services in agriculture. Here, the proper 'conservation-biological-control' could make a difference in agriculture.

Weeds play a pivotal role for competition for nutrients, sunlight (energy) and water. Its outbreaks and resistance to herbicides are also the consequences of faulty agriculture practices. Weed infestation along with the allelopathic effects caused by their (weeds) root exudated to the productive crops causes significant ecosystem disservices in agriculture.

1.6 Messages

Valuation is different than economic price. Economic value should identify all the market, and non-market values (both use and non-use). Often those values are not related to the price which could be directly fetched from the soil as commodity. This is due to 'price generally considers the purchase of limited and/or single uses', while 'economic value' taken into account all possible multiple uses together. Theoretical definitions of price, cost and value are different. The 'price is the amount of money you pay for something'; the 'cost is the price of something that you would be expected to pay'; and the 'value is that quality of an object that permits measurability and therefore comparability' (Robertson 2012). In case of ecosystems, the valuation in an economic term is helpful for comparing one system to other complex one on the basis of their 'socio-ecological' relationships. Edwards-Jones et al. (2000) suggested the ecosystem service values are useful because it provides the information about: (i) importance of ecosystem functioning for human being; (ii) highlights the importance which are of unspectacular, unseen and unattractive ecosystems; (iii) at the local level, it can identify the ESs and help in decision-making; (iv) it could help in understanding the impacts of change and provide important feedbacks to models; (v) communicating a measurable common value for pricing (like US $).

Neo-classic economic value is based on the valuation that human being is generally familiar with which are affecting their everyday lives. Total economic value (TEV) refers to the summation of all relevant non-use and use values provided now and also in the future. It is the sum of both consumer and producer surplus in the demand curve, excluding the cost of production (Costanza et al. 1997a). Within this framework total economic values are divided into two categories, (i) use values and (ii) non-use values.

The use values are typically comprising of three categories: (i) direct use; (ii) indirect use; and (iii) optional. Direct use values include marketable and non-marketable goods and services. These are both the consumptive and non-consumptive use values for goods and services that are locally consumed. Indirect values are those services that nature provides which are not directly consumed. Those are often termed as regulating services. Optional values are those for which people have the option to enjoy those in coming future even not in use currently. Those could be important in the case of land and soil, which generally passed down through the generations. Passive use values are sometimes termed as non-use values. Those are the values that are not associated with actual use, but supportive to some other important services. Similarly, 'bequest value' is given by the people by recognizing the future values of certain commodities or services and those are directly related to the resources going to be used by the future generations (Beaumont et al. 2007).

'Valuation' is the process that quantifies the qualities of certain goods and services at a definite scale. This process is directly linked with ecosystem functions and their management (Costanza et al. 1997b). Regulating services provided by soils have indirect and option-use values for society as well as non-use values relating to the use in future generations. It is the responsibility of the present generation to pass the valuable soil resource in its original potential to the next generation. Therefore, a separate accounting is needed for different services obtained from soil or land resources. In the subsequent chapters, we have given the details of why and how to assign value for different goods and services being extracted from land resources for the purpose of agricultural production.

The concept of ecosystem services in policymaking can be a powerful tool for evaluating different management strategies of natural resources because it gives a logical reason and quantifiable incentive for environmental protection, for biodiversity conservation and for sustainable societal behaviour. Further, the incentives to farmers for providing ES should be based on scale at which a particular ES directly influence the other services. For example, soil fertility restoration, soil retention and pest control through biological predators and pollination act on a farm scale; in that case, direct incentive can be given to farmers who are providing those services. Incentives can be provided either in terms of monetary benefits or providing resources through target-based subsidies. On the other hand, at large landscape scale, like providing enhanced pollinators and predators, groundwater recharging, atmospheric GHGs regulation, maintaining biodiversities and providing aesthetic services depend on economic externalities having lots of non-point beneficiaries. In these cases, designed policy framework is needed to provide incentives to the farmers and other stakeholders who are maintaining those services. Unfortunately, till now not much public policies exist on these aspects which is designed to support the coordinated effort of different stakeholder who are providing those ESs. However, few efforts on this aspect have been initiated through 'Environmental Qualities Incentive Programme' and 'Conservation Security Programme' at state level in the USA (Parkhurst and Shogren 2007; Brander et al. 2012; Estoque and Murayama 2013; Delphin et al. 2016; Fu et al. 2016) that need to be promoted in large scale with site-specific modifications.

References

Assessment MM (2003). Biodiversity and human Well-being: a framework for assessment
Barbier EB (2011) Pricing nature. Annu Rev Resour Econ 3(1):337–353

Beaumont NJ, Austen MC, Atkins JP, Burdon D, Degraer S, Dentinho TP, Derous S, Holm P, Horton T, van Ierland E, Marboe AH, Starkey DJ, Townsend M, Zarzycki T (2007) Identification, definition and quantification of goods and services provided by marine biodiversity: implications for the ecosystem approach. Mar Pollut Bull 54:253–265. https://doi.org/10.1016/j.marpolbul.2006.12.003

Bellamy PH, Loveland PJ, Bradley RI, Lark RM, Kirk GJD (2005) Carbon losses from all soils across England and Wales 1978-2003. Nature 437:245–248

Brander LM, Wagtendonk AJ, Hussain SS, McVittie A, Verburg PH, de Groot RS, van der Ploeg S (2012) Ecosystem service values for mangroves in Southeast Asia: a meta-analysis and value transfer application. Ecosyst Serv 1(1):62–69

Breure AM, De Deyn GB, Dominati E, Eglin T, Hedlund K, Van Orshoven J, Posthuma L (2012) Ecosystem services: a useful concept for soil policy making. Curr Opin Environ Sustain 2012(4):1–8

Bullock JM, Aronson J, Newton AC, Pywell RF, Rey-Benayas JM (2011) Restoration of ecosystem services and biodiversity: conflicts and opportunities. Trends Ecol Evol 26:541–549

Cabral P, Feger C, Levrel H, Chambolle M, Basque D (2016) Assessing the impact of land-cover changes on ecosystem services: a first step toward integrative planning in Bordeaux. France Ecosystem Services 1(22):318–327

Costamagna AC, Landis DA (2006) Predators exert top-down control of soybean aphid across a gradient of agricultural management systems. Ecol Appl 16(4):1619–1628

Costanza R, Cumberland J, Daly H, Goodland R, Norgaard R (1997a) An introduction to ecological economics. St. Lucie Press, Boca Raton, FL

Costanza R, Daly HE (1992) Natural capital and sustainable development. Conserv Biol 6(1):37–46

Costanza R, d'Arge R, De Groot R, Farber S, Grasso M, Hannon B, Limburg K, Naeem S, O'neill RV, Paruelo J, Raskin RG (1997b) The value of the world's ecosystem services and natural capital. Nature 387(6630):253–260

Costanza R, De Groot R, Sutton P, Van der Ploeg S, Anderson SJ, Kubiszewski I, Farber S, Turner RK (2014) Changes in the global value of ecosystem services. Glob Environ Chang 26:152–158

Costanza R, Kubiszewski I, Ervin D, Bluffstone R, Boyd J, Brown D, Chang H, Dujon V, Granek E, Polasky S, Shandas V (2011) Valuing ecological systems and services. F1000 Biology Reports 3

Daily GC, Matson PA, Vitousek PM (1997) Ecosystem services supplied by soil. In: Daily GC (ed) Nature's services: societal dependence on natural ecosystems. Island Press, Washington, DC, pp 113–132

De Deyn GB, Cornelissen JHC, Bardgett RD (2008) Plant functional traits and soil carbon sequestration in contrasting biomes. Ecol Lett 11:516–531

De Deyn GB, Shiel RS, Ostle NJ, McNamara NP, Oakley S, Young I, Freeman C, Fenner N, Quirk H, Bardgett RD (2011) Additional carbon sequestration benefits of grassland diversity restoration. J Appl Ecol 48:600–608

de Groot R (2010) Integrating the ecological and economic dimensions in biodiversity and ecosystem service valuation. In: Kumar P (ed) The economics of ecosystems and biodiversity. Earthscan, London

De Groot R, Brander L, Van Der Ploeg S, Costanza R, Bernard F, Braat L, Christie M, Crossman N, Ghermandi A, Hein L, Hussain S (2012) Global estimates of the value of ecosystems and their services in monetary units. Ecosyst Serv 1(1):50–61

Delphin S, Escobedo FJ, Abd-Elrahman A, Cropper WP (2016) Urbanization as a land use change driver of forest ecosystem services. Land Use Policy 54:188–199

Dominati E, Patterson M, Mackay A (2010) A framework for classifying and quantifying the natural capital and ecosystem services of the soils. Ecol Econ 69:1858–1868

Edwards-Jones G, Davies B, Hussain S (2000) Ecological economics: an introduction. Blackwell Science, Oxford, UK

Ehrlich PR, Mooney HA (1983) Extinction, substitution, and ecosystem services. Bioscience 33(4): 248–254

Estoque RC, Murayama Y (2013) Landscape pattern and ecosystem service value changes: implications for environmental sustainability planning for the rapidly urbanizing summer capital of the Philippines. Landsc Urban Plan 116:60–72

Freeman J, Kolstad CD (2007) Prescriptive environmental regulations versus market-based incentives. Moving to Markets in Environmental Regulation: Lessons from Twenty Years of Experience 3:3–15

Fu B, Li Y, Wang Y, Zhang B, Yin S, Zhu H, Xing Z (2016) Evaluation of ecosystem service value of riparian zone using land use data from 1986 to 2012. Ecol Indic 1(69):873–881

Hewitt A, Dominati E, Webb T, Cuthill T (2015) Soil natural capital quantification by the stock adequancy method. Geoderma 241-242:107–114

Kubiszewski I, Costanza R, Anderson S, Sutton P (2020) The future value of ecosystem services: global scenarios and national implications. Edward Elgar Publishing, In Environmental Assessments

Lal R (2004) Soil carbon sequestration impacts on global climate change and food security. Science 304:1623–1627

Lal R, Follett F, Stewart BA, Kimble JM (2007) Soil carbon sequestration to mitigate climate change and advance food security. Soil Sci 172:943–956

MEA ME (2005 Aug.) Ecosystems and human Well-being: synthesis. Island, Washington, DC

Milder JC, Scherr SJ, Bracer C (2010) Trends and future potential of payment for ecosystem services to alleviate rural poverty in developing countries. Ecol Soc 15(2)

Nayak AK, Shahid M, Nayak AD, Dhal B, Moharana KC, Mondal B, Tripathi R, Mohapatra SD, Bhattacharyya P, Jambhulkar NN, Shukla AK (2019) Assessment of ecosystem services of rice farms in eastern India. Ecol Process 8(1):1–6

Parkhurst GM, Shogren JF (2007) Spatial incentives to coordinate contiguous habitat. Ecol Econ 64(2):344–355

Pimentel D (2006) Soil erosion: a food and environmental threat. Environ Dev Sustain 8:119–137

Power AG (2010) Ecosystem services and agriculture: tradeoffs and synergies. Philos Trans R Soc B: Biol Sci 365:2959–2971

Robertson, M. (2012) Functions, services and values, Wetlandia http://wetlandia.blogspot.co.uk/2012/07/functions-services-and-values.html

Robinson DA, Fraser I, Dominati EJ, Davíðsdóttir B, Jonson JOG, Jones L, Jones SB, Tuller M, Lebron I, Bristow KL et al (2014) On the value of soil resources in the context of natural capital and ecosystem service delivery. Soil Sci Soc Am J 78:685–700

Robinson DA, Hockley N, Cooper DM, Emmett BA, Keith AM, Lebron I, Reynolds B, Tipping E, Tye AM, Watts CW et al (2013) Natural capital and ecosystem services, developing an appropriate soils framework as basis for valuation. Soil Biol Biochem 57:1023–1033

Rutgers M, Van Wijnen HJ, Schouten AJ, Mulder C, Kuiten AM, Brussaard L, Breure AM (2012) A method to assess ecosystem services developed from soil attributes with stakeholders and data of four arable farms. Sci Total Environ 415:39–48

Scharlemann JPW, Tanner EVJ, Hiederer R, Kapos V (2014) Global soil carbon: understanding and managing the largest terrestrial carbon pool. Carbon Manag 5:81–91

Science (2004) Soils—the final frontier. Science 304:1613–1637

SEC (2006) Impact assessment of the thematic strategy on soil protection. Document accompanying, thematic strategy for soil protection. Communication from the commission to the Council, the European Parliament, the European economic and social committee and the committee of the regions. SEC (2006)620. Brussels

Stallman HR (2011) Ecosystem services in agriculture: determining suitability for provision by collective management. Ecol Econ 71:131–139

Sukhdev P, Wittmer H, Schröter-Schlaack C, Nesshöver C, Bishop J, Brink PT, Gundimeda H, Kumar P, Simmons B (2010) The economics of ecosystems and biodiversity: mainstreaming the economics of nature: a synthesis of the approach, conclusions and recommendations of TEEB. UNEP, Ginebra (Suiza)

USDA (2006) USDA roles in market-based environmental stewardship. Departmental regulation 5600–003. Washington, DC

USDA [U.S. Department of Agriculture] (2007) FSA announces nationwide conservation initiative to restore species of concern and their habitats. News release number 1428.07. USDA Farm Service Agency, Washington, D.C., USA

Yu X, Kai A, Gaodi X, Chunxia L (2011) Evaluation of ecosystem services provided by 10 typical rice paddies in China. Journal of Resources and Ecology 2(4):328–337

Zhang B, Li W, Xie G (2010) Ecosystem services research in China: Progress and perspective. Ecol Econ 69(7):1389–1395

Zhang W, Ricketts TH, Kremen C, Carney K, Swinton SM (2007) Ecosystem services and dis-services to agriculture. Ecol Econ 64(2):253–260

Pricing of Agricultural Products, Soil and Water Management

2

2.1 General Methods of Pricing Ecosystem Services in Agriculture

In general, assessment of ES for agriculture could be done by summing up all the individual ES values (including regulatory, supporting and cultural services) (Sandhu et al. 2008; Nayak et al. 2019). That means components of ESs measured for food (p_1), fuel (p_2), fodder (p_3), by-products (p_4), control of pest and diseases (r_1), nitrogen fixation (r_2), carbon flow (r_3), nutrient mineralization (r_4), soil erosion control (r_5), soil formation (s_1), regulation of hydrological flow (s_2), soil fertility maintenance (s_3), cultural (c_1) and recreational services (c_2), etc. are to be numerically summed to get the total value of ESs in agriculture (ES_{total})

$$ES_{\text{total}} = \sum ES_{\text{tangible}} + \sum ES_{\text{non-tangible}}$$

$$\begin{aligned} ES_{\text{total}} &= \sum (p1 + p2 + p3 + p4) \\ &+ \sum (r1 + r2 + r3 + r4 + r5) + (s1 + s2 + s3) + (c1 + c2) \end{aligned}$$

where

ES_{total} = Total ecosystem services for agriculture

ES_{tangible} = Economic value of products and by-products which are traded by farmers and/or stakeholder directly to the market

$ES_{\text{non-tangible}}$ = Services which have economic values by not directly traded in market

$$\begin{aligned} \sum ES_{\text{tangible}} &= (p1 + p2 + p3 + p4) \\ \sum ES_{\text{non-tangible}} &= (r1 + r2 + r3 + r4 + r5) + (s1 + s2 + s3) + c1 + c2 \end{aligned}$$

(Sandhu et al. 2015; Nayak et al. 2019)

P. Bhattacharyya et al., *Pricing of Ecosystem Services in Agriculture: A Basis of Crop Insurance*, https://doi.org/10.1007/978-981-19-4416-1_2

The important points to be noted here that $p_1 \ldots p_4$; $r_1 \ldots \ldots r_5$; $s_1 \ldots s_3$; and $c_1 \ldots c_2$; are not the exclusive lists of components of ESs provided by agriculture. It may be higher in certain cases or may be lower in certain specific situation.

2.2 Pricing of Agricultural Products

The basic provisioning services provided by agriculture through the production of food, fodder, fibre, fuel, by-products, and raw materials for industries which have economic values and can be directly traded in markets by farmers and/or other stakeholders. There are methods for pricing these provisional services. The common method followed to calculate the price of food produced in agriculture by multiplying the minimum support price (MSP) (fixed by the government) in case of India with the whole quantity of produce per annum. This method can be followed to those food product and grains only which have the MSP for a particular year fixed by the Government of the respective countries.

Specifically, the pricing of ESs of food products of major food grains (rice, wheat, maize, millets, pulses, and oilseeds) could be estimated by minimum support price (MSP fixed by government for a particular year) of a particular product with the actual amount of production. The economical values of food/fodder/fibre/by-products whose MSP is not available can be estimated by using farm-gate prices at local markets. For example, price of straw (by product of rice) can be estimated by multiplying total straw yield (approximately 1.5 times of total grain yield) with the local market price of straw, i.e. US$ 15 t^{-1} of straw (Nayak et al. 2019).

$$\begin{aligned}&\text{Pricing of ES of food/fodder/fuel/by products}\\&\quad = \text{Total quantity of product} \times \text{MSP/local market price}\end{aligned}$$

2.3 Valuation of Soil Nutrient Management in Agriculture

2.3.1 Pricing of Nutrient Build-Up

2.3.1.1 Valuation of Soil Nutrients Mineralized

The economic valuation of plant nutrients mineralization can be calculated by converting the total amount of nutrients mineralized to their equivalent price of fertilizers cost. For example, cumulative nitrogen mineralized annually in rice is measured/estimated per hectare, then it is priced/valued at the equivalent price of kg of nitrogen fertilizer (e.g. US$0.0824 kg^{-1}), to get the pricing of nitrogen mineralization. Similar procedure could be followed for the other plant nutrients to get the value of the plant nutrients mineralization as a whole.

Pricing of soil nutrient mineralisation = Quantity of nutrient mineralised
× Corresponding market price of fertilisers containing that particular nutrient

2.3.1.2 Pricing of Nitrogen Fixation

Nitrogen fixation in agriculture either by beneficial nitrogen-fixing microorganism or by leguminous crops is priced by multiplying the total amount of nitrogen fixed per unit land (say 1 hectare or 1 acre) with the actual amount of urea that provide that particular amount of nitrogen and the unit price of urea (US\$0.825 kg^{-1} urea) (Roger et al. 1992; Franco and Balieiro 2000; Nayak et al. 2019). However, pricing may vary from country to country or region to region based on the unit price of urea. The urea subsidy policy of the country has a big role here.

Pricing of nitrogen fixation
= (Total amount of nitrogen − (fixation of nitrogen in soil by biological mass))
×(Specific amount of urea supply that particular amount of nitrogen
× Unit price of urea))
= [Total amount of biological fixed nitrogen × 2.17 x unit price of urea]

2.3.1.3 Methodologies for Pricing Nitrogen Transformation

In agriculture, nitrogen flows taken place among the crops, soils, groundwater, atmosphere and the human society in broader sense. The major four processes involved in the nitrogen transformation among crop system, soil, groundwater and atmosphere are: (i) biological nitrogen fixation in soil; (ii) nitrogen deposition in soil-water system; (iii) ammonia (NH_3) volatilization from soil to atmosphere; (iv) nitrous oxide (N_2O) emission from soil-plant system to atmosphere; and (v) nitrogen losses through irrigation, percolation, infiltration, leaching and drainage. The anthropogenic activities like nitrogen fertilizer application, seeding, harvesting and grain-consumption are also governing the nitrogen dynamics in agriculture and human society. Pricing of nitrogen transformation in agriculture can be done by following empirical functions:

$$A_{ON} = A_P + A_R + A_{NH3} + A_{N2O} + A_D + A_L$$

$$V_{ON} = (A_{ON} \times P_P) + (A_R \times P_N) - [(A_{NH3} + A_D + A_L) \times P_D] - qre \times 320Pg$$

where

A_{ON} = Amount of nitrogen output in agriculture; (kg ha^{-1} $yr.^{-1}$ on nitrogen basis).

A_P = Amount of nitrogen harvested as food/fodder/fuel/fibre; (kg ha^{-1} $yr.^{-1}$ on nitrogen basis).

A_R = Amount of nitrogen harvested as by-product/residues; (kg ha^{-1} $yr.^{-1}$ on nitrogen basis).

A_{NH3} = Amount of nitrogen lost through NH_3 volatilization; (kg ha^{-1} $yr.^{-1}$ on nitrogen basis).

A_{N2O} = Amount of nitrogen lost through N_2O emission; (kg ha^{-1} $yr.^{-1}$ on nitrogen basis).

A_D = Amount of nitrogen lost through drainage; (kg ha^{-1} $yr.^{-1}$ on nitrogen basis).

A_L = Amount of nitrogen lost through leaching; (kg ha^{-1} $yr.^{-1}$ on nitrogen basis).

V_{ON} = Economic value of nitrogen output in agriculture (US\$ ha^{-1} $yr.^{-1}$).

P_P = Replacement of marginal price of food protein (e.g. US \$16.22 kg^{-1}; as pure nitrogen for rice; Gerard 2009; converting the US\$ from 2009 to 2022).

P_N = Replacement price of nitrogen fertilizer (e.g. US \$0.86 kg^{-1}; as pure nitrogen; DPNDRC 2010; converting the US\$ from 2009 to 2022).

P_D = Replacement cost of nitrogen losses through NH_3 volatilization, leaching, N_2O emission (e.g. US \$13.26 kg^{-1} y^{-1}; as pure nitrogen; converting the US\$ from 2009 to 2022).

2.3.2 Pricing of Soil Fertility

Soil fertility service is a major ES provided by agriculture which often ignored or taken to be granted. However, this is one of the major supportive services provided by agriculture. Neglecting its importance and value, the soil could be gradually degraded, and soil health would be deteriorated day by day. Protecting the soil health is equally important as mitigation of climate change. Both are integral part of livelihood and food security. Contribution of essential plant nutrients like nitrogen, phosphorus, potassium, calcium, magnesium, sulphur, boron, iron, manganese, copper, zinc, molybdenum and chlorine provided by soils to the plants must be valued. One of the ways to estimate the economic value of soil nutrient supply is to measure the nutrient uptake by plant from soil and multiplying it to the unit price of respective fertilizers contributing the specific nutrients. The pricing varies with unit price of fertilizers in local markets. The relative contribution of soil to supply the available nutrients to plants/crops depends on fertilizers/nutrients use efficiencies. The tracer technique is one of the widely used method to precise quantification of the nutrients supply of soil to plants.

$$\begin{aligned}&\text{Pricing of soil fertility}\\&= \text{Soil nutrient contribution} \times \text{Unit price of respective fertilizers in local market}\end{aligned}$$

$$\begin{aligned}&\text{Soil nutrient contribution}\\&= \text{Total uptake of nutrients} - \text{Fertilizer nutrient contribution to uptake}\end{aligned}$$

2.4 Pricing of Ecosystem Services in Amelioration of Problem Soils

Problem soils hinder crop growth and yield by restricting supply of water and nutrients from soil environment to plant. Problem soils include sodic soil (alkali soil), saline soil and acid soils. Alkali soil contains enough exchangeable sodium percentage (ESP) (>15%) to cause soil dispersion and increase in soil pH (>8.5), thereby adversely affecting both the physical and nutritional properties of soil and thus significantly reducing crop yield. The electrical conductivity (EC) of soil saturation extract of alkali soil is generally <4.0 dS m^{-1}, whereas sodium absorption ratio (SAR) is >13. Saline soils adversely affect plant growth due to excess presence of neutral soluble salts in the soil systems. Therefore, EC of saturated soil extracts of saline soils are >4 dS m^{-1}; however, soil pH is <8.5, ESP is <15% and SAR is <13. Due to presence of excessive amount of salts, plant roots are unable to absorb water from soil solution even under wet soil condition. This is because of low osmotic potential of soil water due to presence of high concentration of soluble salts as compared to the osmotic potential of plant cell. The saline soil can be grouped in different categories as per the EC of soil saturation extract and feasibility of crop cultivation (Table 2.1).

Acidic soils generally have pH less than 7 and depending upon the pH value, it can be classified into five broad groups: slightly acidic (pH 6.0–6.6), medium acidic (pH 5.6–6.0), strongly acidic (pH 5.1–5.5), very strong acidic (pH 4.6–5.0) and extremely strong acidic (pH <4.5). Acid soils are generally found in high rainfall areas, because of leaching of soluble basic salts such as those of Ca, Mg, K, Na, along with drainage water and insoluble acidic residues mainly composed of oxides and silicates of Fe, Si and Al accumulates in soil matrix in excessive amount. High degree of soil acidity decreases the availability of plant nutrients, particularly P, Ca, Mg, Mo and B and thus plant growth is restricted. Even, the toxicity of Al and Fe in acid soil reduces the soil productivity. Soil acidity below pH 5.5 is injurious to plant roots. In India, about 24.66 million ha is covered under problem soils and among these 6.73 m ha is affected by alkali and saline soil, whereas 17.93 m ha area is affected by acid soil.

Reclamation of alkali soil basically requires partial or complete replacement of Na from exchange site with Ca or Mg followed by removal of the Na salt with leaching water. Gypsum (calcium sulphate, $CaSO_4$) is by far the most used soil amendment material for replacement of Na ion from soil exchange. Other soil

Table 2.1 Soil salinity classes and feasibility of crop growth

Salinity class	EC (dS m^{-1})	Effect on crop plants
Non-saline	0–2	Salinity effects negligible
Slightly saline	2–4	Yields of sensitive crops may be restricted
Moderately saline	4–8	Yields of many crops are restricted
Strongly saline	8–16	Only tolerant crops yield satisfactorily
Very strongly saline	>16	Only a few very tolerant crops yield satisfactorily

amendment materials used for reclamation of alkali soil are sulphur, sulphuric acid, calcium chloride, aluminium sulphate, etc.; however, these are expensive and therefore uneconomic to use. Press mud from sugar factories can also be used as soil amendment. Iron pyrite is another promising reclamation option provided it contains a minimum 8% water soluble sulphur. Proper land levelling with strong field bunds at all sides of the field is a must for initiating the reclamation. The amount of soil amendment is dependent on initial value of ESP and to be obtained through laboratory analysis of soil sample. Gypsum requirement (GR) is calculated based on the following calculation:

$$GR\ (\text{t/ha}) = \frac{(\text{Initial ESP} - \text{desired ESP})}{100} \times \text{CEC}(\text{meq}/100g) \times 86 \times \frac{10}{1000} \times 2.25 \times \frac{1}{\text{purity}}$$

For example, in case the ESP from an initial value of 25% is desired to be reduced to 5% with a soil CEC of 20 meq/100 g, then the GR is 7.74 t ha^{-1}, provided the purity of gypsum is 100%. Reclamation of saline soil is done through leaching the salts below the soil profile followed by drainage. Otherwise, salt tolerant crops, e.g. sorghum, wheat, oats, barley, pomegranate, date palm, etc. are suggested to grow. Hybrid approach of adopting the abovesaid two approaches for saline soil reclamation is most preferred. Amelioration of acid soil is done through lime application in field. Lime requirement depends on the soil texture, initial soil pH, target soil pH and quality of liming material. For practical purpose, the target pH of soil should be above 6.0. As per the prescribed tabulated chart, requirement of $CaCO_3$ is identified as per the current soil pH in Shoemaker, McLean and Pratt (SMP) buffer solution and target soil pH followed by calculation of lime requirement considering calcium carbonate equivalent of lime (Cao) as 179%. However, lime requirement at any case should not be higher than 3 t ha^{-1}, otherwise it can upset soil nutrient supply.

Reclamation of problem soil directly increases the crop yield, monetary value of which can be accounted as the provisioning service of soil reclamation as follows:

$$\text{Provisioning service}\ (\ \text{ha}^{-1}) = \Delta \text{Y}_{\text{crop}}(\text{t ha}^{-1}) \times \text{Price of crop}\ (\ \text{t}^{-1})$$

where ΔY_{crop} is the incremental crop yield due to soil reclamation. For example, Kumar et al. (2012) reported an increase of maize yield by 32.4% by application of lime @ 3 t ha-1 in acid Alfisol of Umiam, Meghalaya in north-east India. Considering the minimum support price of maize in India during 2020–2021 as INR 18500 t^{-1} ($ 245 t^{-1}; $ 1 = INR 75.48) and average maize yield of 1.5 t ha^{-1} in Meghalaya, the provisioning service of liming in north-east India is calculated as about $ 187 ha^{-1}. Kumar et al. (2017a) also reported the increase in sesamum yield from 183 kg ha^{-1} to 292 kg ha^{-1} (~60% increase) after application of lime @ 350 kg ha^{-1}. Likewise, the provisioning services of reclaiming saline and sodic soil can also be calculated. It is to be noted here, the cost of reclamation, the transportation cost of material required for reclamation, application cost, etc. need to be accounted for calculation of net provisioning service.

For example, application of gypsum was also found to increase wheat yield in Typic Haplocambids soils of western Rajasthan (Kumar et al. 2016). The availability of P, Fe and Zn in rhizosphere was increased by 25.91, 32.72 and 38.46%, respectively over control with the application of gypsum @100% of gypsum requirement (GR). Application of gypsum also found to increase plant height, tillers per plant, leaves per plant and ear length of wheat by 13.16, 22.98, 23.12 and 13.82%, respectively over control. The grain and straw yields increased by 39.72 and 37.58% after gypsum application (Kumar et al. 2017b).

In addition to direct impact of soil reclamation on crop yield, it also improves the availability of major soil nutrients, improves the biological productivity of soil, improves the soil organic carbon stock, etc. Increase in nutrient availability, SOC status and biological activity directly reduces the cost incurred for fertilizer application, and therefore can reduce the cost of cultivation. All these services of soil reclamation can be quantified as supporting services as follows:

$$\text{Supporting service}\ (\text{ha}^{-1}) = \sum_{i=1}^{n} \Delta N_i\ (\text{kg ha}^{-1}) \times C_i\ (\text{kg}^{-1})$$

where ΔN_i is the change in the ith nutrient application rate after soil reclamation and C_i is the unit cost of the ith nutrient. For example, Inagaki et al. (2016) reported the increase in biological activity, carbon pools and agronomic productivity of highly weathered soil in Brazil.

2.5 Valuation of Water Management in Agriculture

There are debate between the extent of natural and managed landscape or ecosystems in present day situation. In real sense, 90% of earth ecosystems are influenced by the anthropogenic (human induced) activities including food production, forage, hunting, timber gathering, water harvesting, touring, recreational, etc. And there would be no natural ecosystem left, if we consider the climate change influences. So, every landscape or ecosystem including water cycles are some of the either way managed by human being. Therefore, the practical question is what beneficial services we want to provide by water managed (landscape managed) and how we can manage the systems to get desired services on sustainable basis.

Water is a single commodity or association with other natural resources provides highest provisional ESs to human being. Those service can be listed as drinking water, hydro power generation, all type food (agriculture, livestock, rearing, wild animals, fisheries, etc.). It provides regulatory services in terms of climate regulation, hydrological flow maintenance, erosion control, hazard regulation, etc., therefore costing of any infrastructure or operation which is needed for treating water to make it suitable for use (original form) can be used for valuation (pricing) of ESs provided by water for regulatory services.

The payment of ecosystem services (PES) can be defined as the direct or indirect payment made by the beneficiaries to the provider for sustaining the desired level of that services (UNDP). The ESs provided by the 'water supply management system' (WSMS) is huge. The primary hydraulic infrastructure in the ecosystem helps in storing and distribution of the valuable water providing social, economic and cultural benefits to the society. Specifically, 'agriculture water control structure' (AWCS) additionally provide irrigation and drainage services, it also helps in aquifer recharging, regulation of baseflow and controlling water erosion. Further, well-maintained 'rice paddy' field (itself a definite hydraulic structure) provides bundle of associated ESs, like modification of hydrological properties of soil, flood control, groundwater recharging and controlling soil erosion.

Those AWCS and or 'water supply management' systems need continuous financial assistance to perform their function effectively. It requires proper long-term financing. But unfortunately, most of the cases it is missing. Often the farmers are alone who are the sole contributor of paying water user fees, which is not sufficient to maintain the systems at desired level. There are limitations of funding for 'water management systems' due to policy ambiguity and more so for the non-counting of the 'social value' of irrigation water to its 'economic value'. So, alternative sources of funding need to be explored for proper function of AWCS or 'water supply management systems'. In this connection, PES can play a vital role for proper functioning 'water supply management systems'. The cost burden could be shared by various beneficiaries of ESs. For example, for providing the ES like recharging the groundwater by flooding rice paddy in mid-catchment, the farmers of the mid-reached should be considered as seller. Whereas certain company like, 'Sony Semi-conductor Co. Kumamoto Technology Centre' should be buyer of the water at down-command area. Here, the acreage of flooded paddy fields can be taken for evaluation of ES. Similarly, the cattle ranches could be the seller for providing the amount nutrients that can be removed (or amount of storm water impounded), while 'the South Florida Water Management District' can be the buyer. Here the ES evaluation was done through modelling to estimate flow of ESs. The ES, like improvement of water quality by reducing the sediment load of the river at upper-reaches of the watershed can be sold by the upstream-farmers. Certain companies at lower-reached (say 'Coca Cola Kwanza Ltd. and DAWASCO') can act as buyer of that ES. Here also, the acreage of recommended land management practices followed could be used for evaluation of ES (Okiria et al. 2021).

As there is no conventional market of non-tangible ES, so approach of PES may be started in project mode funded by government or non-governmental organizations. The link between water supply management practices and ESs should be established at the beginning of the project implementation period. During the project implementation, monitoring of water quality and hydrology should be done for pricing of ESs flows. Here, hydrological modelling can play a significant role for precise estimation of marginal ES flows for adopting certain water-supply management practices. The focus of PES project should be on securing the continuous flow of ESs, not on complementary financing of the whole irrigation project. Proper planning of PES project for 'water supply management systems' can be a game-changer in financing the irrigation-drainage schemes. It would help in sustainable maintenance of the infrastructure and operation expenditure. However, proper book-keeping for contribution of farmers towards providing the ESs must be ensured on equitable basis and negotiation of buyers and sellers can be done through well-organized farmers' union.

In this connection the following approach can be followed for pricing of ESs provided by water management which could be partially addressed through the payment of water supply.

$$\text{Rate tariff of payment of water supply(US \$ per ha)} \\ = \frac{\text{Expenses of water supply organization for rendering the services(US \$)}}{\text{Quantity of water used(either in ha of irrigated area or volume of water used, in m}^3\text{)}}$$

$$P_{TW} = \frac{C_{WSS}(US\$)}{Q_{WU}(ha\ or\ m^3)}$$

$$C_{wss} = (C_{GWSS} - S_{WSS} + C_{PWSS}) \times \text{Value added taxes}$$

where

P_{TW} = Rate tariff of payment of water supply (US \$ha^{-1}).

C_{WSS} = Cost/expenses of water supply organization for rendering the services (US \$).

Q_{WU} = Quantity of water used (hectare (ha) of irrigated area or volume of water used for irrigation (m^3)).

C_{GWSS} = General cost of water supply including irrigation line layout cost, distribution cost, operational cost and overhead cost.

S_{WSS} = Subsidy provided by the state or country to the farmer.

C_{PWSS} = Expenses of property (farm assets) used for supplying the water services.

In general, there are no standard method available for payment of water supply to irrigation system. Even the above-mentioned approach can be used. However, the efficiency of water uses, and frequencies of irrigation also need to be assessed simultaneously for valuation.

The monitoring and measurement of water quality paraments are necessary for transparency of the system. Advanced technologies including remote sensing, drone-based monitoring and modelling should be followed for reducing the cost of PES-based 'water supply management' projects. Studying the linkage between adopted/recommended water-management practices and their marginal contribution towards the flow of ecosystem services is the researchable issue. In this connection, Okiria et al. (2021) proposed a thematic project of 'Developing PES schemes for the operation of agricultural water control infrastructure as a complementary finance mechanism for operation, maintenance and repair' could be a viable alternative. The schemes/project must give greater importance to the opportunity cost which is the backbone of the sustainability of the scheme. The project should use PES as the 'sustainable irrigated agriculture policy instrument' rather than only the scheme of upstream forest-catchment management to preserve and/or improve water quality and quantity. The project should also consider the range of stakeholders enjoying the ESs and the socio-economic condition. Moreover, training and capacity building of the farmers should be the part of the projects. The long-term participation of the stakeholders is important that can be achieved through adopting the participatory approach between sellers, buyers and other stakeholders.

2.5.1 Flood Control and Water Regulation

Considering irrigation, drainage, groundwater recharging through leaching process for surface and groundwater regulation and contribution of ridges/bunds to flood control, the pricing of these services can be done by following the different functions.

2.6 Future Road Map

2.6.1 Option of Markets for Ecosystem Services in Agriculture

Creation of markets, government regulation and/or private action, conservation programmes are the ways by which resources can be allocated to agricultural practices that provided non-tangible ecosystem services. In practical sense, market-based approaches are found more effective than government regulation as beneficiaries directly pay to the service provider. However, creating markets for 'social-good' type of ESs is difficult and more challenging to allocate resources for those services. Some successful approach in these regards were eco-labelling, water quality trading, GHGs trading, carbon trading, wetland-mitigation and 'fee-hunting' (Brown and Reeder 2007; Ribaudo et al. 2010). But success of these approaches depends on how precisely 'offsets' are verified and enforced. All these trade programmes must ensure that agricultural sources should allow to produce and sell the 'offsets'. The 'offsets' should be real, cheaper than emission reduction costs fixed by regulated authorities. Transaction cost between buyers and sellers also must be kept at low level. The important point needs to be remembered here that the 'offsets' should be interchangeable, compliance-flexible and an integral part of regulation guidelines. For example, in 'GHGs trading' approach, the markable 'offsets' are carbon sequestration in soil, methane destruction, and conversion of cropland to grassland or forest. In case of 'eco-labelling' approach of organic produce has a well-established market but completely depends on consumer choice. Here, also certification of 'organic product' is an important and challenging issue.

The basic purpose to value ESs in agriculture is to encourage or motivate farmers or land managers to maintain ESs at a desired level for human well-being. Lack of markets of most of the ESs in agriculture is the key problem. In true sense market does not exist to many supportive, regulatory and cultural services in agriculture. The price of ES also varies between farmers (producers) and beneficiaries (consumers). In those cases, the approach of willingness to provide certain ES in exchange of specific compensation for farmer can be adopted ('willing to accept' (WTA)) (Zhang et al. 2007; Swinton et al. 2007; Sandhu et al. 2008; Porter et al. 2009; Ribaudo et al. 2010; Stallman 2011). In other side 'willingness to pay' (WPT) by consumers for getting satisfactory ES needs to be considered simultaneously. Few methods documented by researchers to calculate the market value of 'consumer WPT and producer WTA' are briefly discussed below:

1. Estimating the travel cost to reach the outdoor site of recreation can be the price of 'recreational-ES' provided by agriculture and it is linked with 'consumer WTP', e.g. fishing, hunting, creating greeneries.
2. The other approach to fix the passive value of certain ES which people usually hold whether they directly use them. It is termed as 'contingent valuation' which could be estimated through people-survey and based on the 'consumer WTP' and 'producer WTA' (Freeman 2003; Colombo et al. 2006). The ESs provided by agriculture like wildlife habitat, water quality maintenance, scenic-beauty for refreshment and pure oxygen, etc. can be priced with this method.
3. Pricing the ESs based on the relationship between land characteristic and land-property price is called 'hedonic-variation'. Here, ES linked with the price of farmland (land-property) which also directly or indirectly affected by neighbouring farms (surrounding land properties). Neighbouring farms exert both positive and negative influences on land in question. For example, waste disposal of residential colony negatively affects the farmland productivity. On the contrary, certain farm that rich in beneficial soil-microbes can provide higher crop yield should be priced higher because it provides higher provisional, supportive and regulatory services simultaneously for long run. But in this method of valuation, market participation and proper knowledge of both buyers and sellers about how the ES influences in crop productivity and linkage between ESs and agricultural earning are necessary.
4. Replacement and mitigation cost-based approach is also an effective way to price ESs in agriculture and create market for those. For example, value of additional fertilizer input required for maintaining soil fertility; additional cost of dredging water bodies if soil erosion and sedimentation are not checked by conservation farming can be alternatively put as price for ESs provided by agriculture.
5. Another approach could be 'production-function' estimation. Here, different production function (inputs at different level, e.g. food, water, energy) is related to agricultural outputs (say yield) at various input level either single or in combination. The basic principle is to compute, how the change in ES affects the expected present value of profits. In other words, the base yield (without any external output) is compared with respective inputs (production function) separately and/or collectively (Dillon and Anderson 2012; Liu et al. 2006).
6. The well-being of consumer can be quantified in terms of change in the value of ES. In this case, even a small price of ES could be a large economic value when we consider a large number of consumer affected/influenced by those.

2.7 Messages

Valuation and/or pricing of ESs for agricultural products and soil-water management should not be done in isolation. Soils and water provide necessary supporting and regulatory services for crop production through nutrients and soil moisture. On the other hand, residues and by-products of the economic crop add necessary organic matter to soil, which helps in maintaining soil fertility. Plant roots hold soils and control erosion. Roots and other rhizospheric organism help in soil aggregation and enhance water holding capacity. Plant-soil-microbial continuum helps in CO_2-O_2 balance in earth atmosphere through photosynthesis, respiration and decomposition processes. Therefore, provisioning, and regulatory/supporting services are interlinked in agriculture. They must be valued together to avoid double counting. Here, human resources and beneficial management practices should also consider.

References

Brown DM, Reeder RJ (2007) Farm-based recreation: a statistical profile. Economic research report 53. U.S. Department of Agriculture, Economic Research Service

Colombo S, Calatrava-Requena J, Hanley N (2006) Analysing the social benefits of soil conservation measures using stated preference methods. Ecol Econ 58(4):850–861

Dillon JL, Anderson JR. The analysis of response in crop and livestock production. Elsevier; 2012

DPNDRC (2010) Compilation of cost and benefit data of agricultural products of China in 2010. Chinese Statistical Press, Beijing

Franco AA, Balieiro FDC (2000) The role of biological nitrogen fixation in land reclamation, agroecology and sustainability of tropical agriculture. In: Transition to global sustainability: the contribution of Brazilian science, pp 209–233

Freeman AM (2003) VThe measurement of environmental and resource values. Resources for the Future, Washington DC

Gerard LW (2009) Rebuild America Clean and Green. NEW SOLUTIONS: J Environ Occup Health Policy 19(2):149–156

Inagaki TM, Moraes Sa JC, Caires EF, Goncalves DRP (2016) Lime and gypsum application increases biological activity, carbon pools, and agronomic productivity in highly weathered soil. Agric Ecosyst Environ 231:156–165

Kumar M, Hazarika S, Choudhury BU, Ramesh T, Verma BC, Bordoloi LJ (2012) Liming and integrated nutrient management for enhancing maize productivity of acidic soils of Northeast India. Ind J Hill Farming 25(1):35–37

Kumar S, Meena RS, Yadav GS, Pandey A (2017a) Response of sesamum (Sesamum indicum L.) to Sulphur and lime application under soil acidity. Int J Plant Soil Sci 14(4):1–9

Kumar M, Singh R, Kar A (2016) Interventions.of high residual sodium carbonate water-degraded soils amelioration technology in Indian Thar Desert and farmers' response. Nat Sci Acad Letter 39(4):245–249

Kumar M, Singh R, Kumawat RN (2017b) On–farm performance of wheat grown on sodic soils irrigated with high RSC water in arid regions of Rajasthan. J Soil Water Conserv 16(3):279–284

Liu Y, Swinton SM, Miller NR (2006) Is site-specific yield response consistent over time? Does it pay? Am J Agric Econ 88(2):471–483

Nayak AK, Shahid M, Nayak AD, Dhal B, Moharana KC, Mondal B, Tripathi R, Mohapatra SD, Bhattacharyya P, Jambhulkar NN, Shukla AK (2019) Assessment of ecosystem services of rice farms in eastern India. Ecol Process 8(1):1–6

Okiria E, Zaki MK, Noda K (2021) A review of payment for ecosystem services (PES) in agricultural water: are PES from the operation of agricultural water control structures ubiquitous? Sustainability 13(22):12624

Porter J, Costanza R, Sandhu H, Sigsgaard L, Wratten S (2009) The value of producing food, energy, and ecosystem services within an agro-ecosystem. AMBIO J Hum Environ 38(4): 186–193

Ribaudo M, Greene C, Hansen L, Hellerstein D (2010) Ecosystem services from agriculture: steps for expanding markets. Ecol Econ 69(11):2085–2092

Roger PA, Ladha JK, Biological N (1992) 2 fixation in wetland rice fields: estimation and contribution to nitrogen balance. In: Biological nitrogen fixation for sustainable agriculture. Springer, Dordrecht, pp 41–55

Sandhu H, Wratten S, Costanza R, Pretty J, Porter JR, Reganold J (2015) Significance and value of non-traded ecosystem services on farmland. PeerJ 3:e762

Sandhu HS, Wratten SD, Cullen R, Case B (2008) The future of farming: the value of ecosystem services in conventional and organic arable land. An experimental approach. Ecol Econ 64(4): 835–848

Stallman HR (2011) Ecosystem services in agriculture: determining suitability for provision by collective management. Ecol Econ 71:131–139

Swinton SM, Lupi F, Robertson GP, Hamilton SK (2007) Ecosystem services and agriculture: cultivating agricultural ecosystems for diverse benefits. Ecol Econ 64(2):245–252

Zhang W, Ricketts TH, Kremen C, Carney K, Swinton SM (2007) Ecosystem services and dis-services to agriculture. Ecol Econ 64(2):253–260

Pricing of Soil and Water Conservation in Agriculture

3

3.1 Pricing of Soil Erosion Control, and Soil Build-Up

Various soil processes are responsible for the flow and maintenance of ecosystems services within a soil type in a landscape. Soil degradation alters or transforms these flows in the domain of soil critical zone (the zone where pedosphere interacts with lithosphere, hydrosphere, biosphere and atmosphere). The major drivers contributing to soil degradation are anthropogenic pressures from land use and land management practices. The human-induced pressure causes land degradation like soil erosion, depletion of soil organic matter, loss of soil biodiversity, acidification, mechanical compaction, secondary salinization (from irrigation water) and contamination by industrial wastes (EC 2006; FAO and ITPS 2015). These degradations are damaging the capacity of soil to perform optimum soil functions.

Soil erosion is the major cause of land degradation. Soil erosion is not only an environmental issue; it also causes huge losses to the economy through loss of provisioning, regulating and supporting ecosystem services. Study revealed that global economic loss from soil erosion is around US$ 8 billion annually, due to reduced soil fertility, decreased crop yields and increased water usage (Hussein 2008). However, proper implementation of soil and water conservation (SWC) measures and sediment control programmes can reduce the erosion and sediment loads of water bodies since majority of soil and water conservation programmes are based on watershed approach and which are being increasingly adopted worldwide; therefore, the impact assessment of watershed management programmes on the erosion and production process is of immense importance. However, such information is not readily available on a regional scale. Although many literatures provide examples of plot and small catchment experiments that only demonstrate the success of soil and water conservation measures and improved management practices in reducing soil erosion and water loss, there are limited quantitative evidence available on the impact of such measures in improving the ecosystems services.

P. Bhattacharyya et al., *Pricing of Ecosystem Services in Agriculture: A Basis of Crop Insurance*, https://doi.org/10.1007/978-981-19-4416-1_3

Loss of biomass, one of the major consequences of soil erosion, has serious negative impacts on food security, livelihood and environment at local, regional and national scale. For example, major *rainfed* crops in India annually experience a loss of 13.5 million tonnes due to water erosion which amounts to a loss of INR ₹ 205 billion in monetary terms (Sharda and Dogra 2013). Apart from economic loss, erosion transported toxic chemicals that pollute water reservoirs, streams and groundwater; and degrade soil health. Moreover, release of additional carbon dioxide into the atmosphere through organic matter dislodgement followed by decomposition has a serious concern in climate change perspective (Mandal et al. 2020). Erosion also weakens soil's condition in providing a favourable habitat for soil biodiversity. Therefore, while assessing the loss of ecosystem services due to soil erosion, we should consider both the on-site and off-site damages. On-site crop productivity loss is just one indicator of provisioning services. Further, during estimation of losses of provisioning services the economic losses of valuable above-ground biomass (i.e. straw/stover/fodder/fibre) should be included along with the price losses of grains. Similarly, soil erosion at the 'on-site' causes loss of soil organic carbon (SOC), nutrients and available soil water which are the loss to supporting and regulating services.

Valuation of ecosystem services in India has been generally estimated by using the approach of replacement cost of the major nutrients (e.g. organic carbon (C), nitrogen (N), phosphorus (P), and potassium (K), sulphur (S)). Generally, the replenishing cost of the loss of major nutrients at 1 mm-ha topsoil (i.e. 1 mm topsoil erosion occurs from 1 ha land) is considered. Erosion from this magnitude (1 mm-ha topsoil) results a loss of 150–225 kg C, 12.97–28.26 kg N, 1.14–6.79 kg P, 13.03–25.5 kg K and 3.74–4.85 kg S (Mandal et al. 2021). Now, to substitute these nutrients losses, there would be additional requirement of 0.67–1 tonne compost, about 21.02–50.56 kg urea, 11.50–68.51 kg single superphosphate (SSP), 18.21–39.42 kg muriate of potash (MOP) and about 3.42 kg of elemental sulphur (90% S). The economic cost to replace the lost nutrients through fertilizers and manure is about INR ₹ 2200–3742 per hectare per year as per 2020–2021 unsubsidized price (Table 3.1). Though visual perception is insufficient to judge the loss of 1 mm topsoil, it is equivalent to 15 tons per hectare (assuming bulk density of 1.5 g cm^{-3}). The total annual soil loss in India is up to 5.1 billion tons with an average 15.35 tons per hectare in India (Sharda and Ojasvi 2016). Therefore, to recover the total major nutrient losses due to soil erosion an expenditure of INR ₹ 0.73 to 1.27 trillion to be incurred annually and that is not usually taken into consideration while assessing the cost of soil erosion. The entire nutrient displaced through erosion does not lose its total functionality as there is a substantial recovery of ecosystem services of the displaced nutrients from the flood-prone area and aquatic production system. Sharda and Ojasvi (2016) reported that out of the total gross soils eroded in India, about 34.1% is deposited in the reservoirs, 22.9% is discharged outside the country (mainly to oceans) and 43.0% is displaced within the mainland. The sediment going out of the landmass of the country is primarily the finer particles and rich in nutrients. Moreover, the sedimentation in different

Table 3.1 Quantity and valuation of nutrients lost through 1 mm of topsoil erosion from 1 hectare land

Nutrient losses (kg)	Prime replacement materials (kg)	Replacement cost-unsubsidized INR ₹ (Rs.)
Carbon: 150–225	FYM: 660–1000	INR ₹ 1340–2000 US$ 19.1–28.6
Nitrogen: 12.97–28.26	Urea: 21.02–50.56	INR ₹ 295–710 US$ 4.2–10.1
Phosphorus:1.14–6.79	Single super phosphate: 11.50–68.51	INR ₹ 38.27–228 US$ 0.6–3.2
Potassium: 13.03–25.5	18.21–39.42	INR ₹ 381–803 US$ 5.4–11.5
Sulphur: 3.74–4.85	Elemental sulphur: 3.42 (purity 90%)	INR ₹ 146 US$ 2.1
Total replacement cost (INR ₹ ha^{-1}; US$)		**INR ₹ 2200–3742** **US$ 31.42–53.5**

Note: 1 US$ = 70 INR ₹, FYM = Farm Yard Manure

reservoirs reduces the carrying capacity for water which otherwise would have been used for irrigation purpose.

3.2 Pricing the Ecosystem Benefits of Soil Conservation Measures

Soil conservation measures are sustainable if the supporting, provisioning, regulating and cultural ecosystem services provided by soil are maintained or enhanced without significantly impairing the soil functions. Maintaining balance between the supporting and provisioning services like regulating hydrological flow, atmospheric gas regulation, soil fertility maintenance and crop production are important. The implementation of soil management practices is typically made at local level having widely different socio-economic contexts. Site-specific soil conservation measures have been implemented for last 4–5 decades in different countries at individual, community and watershed level with the aid from government and non-governmental organizations.

A land use or land management to maximize a specific ES (i.e. food production from terraced cropland) may reduce the other soil functions and hence the other associated ESs. In the hill and mountain ecosystem, crop production has direct linkage to surrounding ecosystems. While the most important concern is to protect soil productivity and sustain production, an equilibrium should be maintained between agricultural production and surrounding natural ecosystems (Lefroy et al. 2000; Saxena et al. 2005; Mandal et al. 2010;). Failing to stop the continued land degradation especially in Himalayan ecosystems would adversely affect several ecosystem functions, not only in the highlands, but also in the downstream command areas (Sharma 2004; Mandal et al. 2010). For easy understanding, some important soil functions are compared for two contrasting ecosystems at three different

Table 3.2 Comparison of multiple soil functions by landscape and land use in hill and mountain ecosystem (adopted from Mandal et al. 2010 and recalculated based on analytical hierarchy process)

Sl. No.	Soil quality functions	Landscape					
		Top-hill		Mid-hill		Lower-valley	
		Forest land use	Terraced agriculture (*rainfed*)	Forest land use	Terraced agriculture	Forest land use	Terraced agriculture (irrigated)
1.	Infiltration of water and air	0.76	0.61 *(−) 19.73%*	0.58	0.41 *(−)29.3%*	0.67	0.55 *(−) 17.9%*
2.	Retention/ conservation of soil moisture	0.64	0.39 *(−) 39.06%*	0.58	0.41 *(−) 29.31%*	0.61	0.54 *(−) 11.47%*
3.	Organic matter and nutrient cycling	0.94	0.65 *(−) 30.85%*	0.88	0.56 *(−) 36.36%*	0.86	0.66 *(−) 23.25%*
4.	Resistance to erosion	0.64	0.63 *(−)1.56%*	0.50	0.40 *(−)20.0%*	0.63	0.58 *(−)0.79%*
5.	Habitat for soil organisms.	0.87	0.40 *(−) 49.38%*	0.65	0.45 *(−) 30.76%*	0.76	0.63 *(−) 17.10%*
Overall soil quality		0.77	0.55	0.64	0.43	0.71	0.59

Parentheses include percentage change with respect to corresponding native sites at each location

locations within a landscape in hill and mountain ecosystem, India (Table 3.2). Overall, the aggregated soil quality was better in adjoining forest sites than that of terraced croplands. It clearly showed that to maximize the crop productivity through intensive cultivation, there is massive compromise with the soil ecosystem functions with regard to organic matter supply and nutrient cycling and habitat for flora and fauna. To improve land use decisions, value of ecosystem services is needed for understanding the impacts of various management practices at multiple scales. Although, the land use practices in upper-slope and mid-slope locations were more or less the same, the magnitude and direction of change of some of the soil functions were dissimilar. This may be due to different soil management techniques which had helped to reduce the erosion and improved soil conditions particularly in upper and mid-slope regions. Therefore, from soil quality aspect, it could be concluded that ecologically applicable destruction took place in *rainfed* old croplands within mid-slope and high-slope areas. Compared to the adjoining reference sites (forest) there is degradation of one or many soil functions due to intensification. The degree of degradation is called 'eco-deficit'. Optimization of 'eco-deficit' rather than maximizing one or few ESs should be the major objective of any soil and water conservation programme. The continued degradation of the fragile Himalayan region would affect adversely the socio-economic and environmental stability of the region. A credible measure of 'eco-deficit' by assigning pricing (monetary value;

footprint) of each land use and land management practices should be made compulsory in any land-based decision-making.

A set of suitable indicators need to be developed to analyse the impact of watershed management on various ecosystems services. The indicators should be scientifically robust to essentially assess the measurable impact of different soil and water conservation (SWC) interventions (agronomical, vegetative and engineering measures) in improving the various ecosystem functions and services, socio-economic condition of the community and ensuring ecological strength in the long run.

Steps for assessing soil ecosystem services generated by soil and water conservation practices:

1. Identify the relevant indicators for each ES (e.g. biomass and grain yield for provisioning services; soil retention or regeneration rates and soil fertility status for supporting services; soil water intake and soil carbon status for regulating service; and number of tourists or visitors visited the site for cultural service).
2. Delineate the geo-spatial dimension of the area of concern (e.g. area affected by erosion or soil improvement area due to SWC interventions).
3. Monitoring and recording the data or collect the relevant information pertaining to the respective ES indicators.
4. Select the appropriate method for assigning the monetary value (e.g. fertilizer replacement cost for lost nutrients or soil fertility assessment; cost of irrigation for assigning value to soil moisture conservation; carbon credit for sequestration of soil organic carbon).
5. Sum up of all values of ESs.

A brief account of selection and valuation of individual indicator for each ES is presented below (Table 3.3).

In another study, the ecosystem benefits of different soil and water conservation measures (SWCM) and sustainable land management practices (SLMP) were computed by collecting the research findings carried out at hill and mountain ecosystems of India. The implementation of soil management practices was executed at local level. While identifying and selecting the sustainable land management practices, the major emphasis was given on the agronomic and vegetative measures rather than mechanical measures. From the ecosystem service point of view, they have much better impacts. We considered the mid-term to long-term experiments where the data for at least 3-year period was reported. All together we considered 43 experiments from various hill and mountain ecosystem. The valuation of provisioning service was computed by taking the data on increase in yield of crops, fodder, and above-ground biomass due to imposition of SLMPs. Difference in yield compared to without SLMPs was then multiplied with minimum support price of the crops or the actual market price value of the biomass/straw/stalk. Total value of provisioning services was assigned by taking summation of all values of grain yield and above-ground biomass. Similarly, supporting services were computed by assigning the value to reduction in soil loss in different SLMPs compared to without

Table 3.3 Selection and valuation of individual indicator for each ES provided by soil and water conservation practices

(A) Ecosystem service:	Supporting service (reduction in soil loss/soil retention)
(i) Landscape functions affected by soil-water conservation interventions:	Surface run-off (% of annual rainfall or mm-ha of water), soil erosion rates (Mg ha^{-1} year^{-1}).
(ii) Monitoring indicator:	Annual change of run-off and soil loss values
(iii) Rationale:	The displaced soil material accumulates in channels, streams and rivers, which reduces their storage capacity and increases inundation area, which in turn affects the adjoining crop lands, meandering of river courses, and damage vegetation. Further, soil loss also carries away the nutrient in topsoil layers affecting the productivity.
(iv) Data requirement:	Soil erosion from different land uses, catchment area, nutrient content of soil, market prices of nutrients (cost of fertilizers)
(v) Calculation of the indicator:	Soil loss at the landscape positions and land use systems need to be measured. Difference in volume of soil loss, beyond soil tolerance limit of the region (Mandal et al. 2006, 2010; Mandal and Sharda 2011), comparing after catchment treatment over before imposing SWC measures or reference (control) condition may be worked out.
(vi) Geo-spatial scale:	Specify the area (plots/field, sub-catchment, basin, terraced cropland, agroforestry system, grassland, fallow land and where SWC measures were imposed)
(vii) Methodology:	Soil loss at the catchment outlet is usually measured in watershed. For estimation, one can use the universal soil loss equation (USLE) proposed by Wischmeier and Smith in 1978 which is the most widely used model in predicting the loss of soil. It is described by the following equation, $SL = R \times K \times L \times S \times C \times P$ SL = estimated average soil loss in (Mg ha^{-1} year^{-1}), R = erosivity of rainfall (MJ mm ha^{-1} h^{-1}), K = soil erodibility factor (Mg ha h ha^{-1} MJ^{-1} mm^{-1}), LS = topographic factor integrating slope length and steepness (LS) dimensionless, C = cover-management factor, dimensionless, P = support practice factor, dimensionless $SL_{av} = \frac{\sum_{i=1}^{n} SL_i A_i}{\sum_{i=1}^{n} A_i}$ SL_{av} = average soil loss in the watershed (Mg ha^{-1} year^{-1}), SL_i = average soil loss in the ith land use in the watershed (Mg ha^{-1} year^{-1}), A_i = area under ith land use in the watershed (ha) [Using appropriate analytical method, soil carbon,

(continued)

Table 3.3 (continued)

(A) Ecosystem service:	Supporting service (reduction in soil loss/soil retention)
	nitrogen, phosphorus and potash stocks, in laboratory, can be determined from the representative run-off sample which can be then extrapolated for total annual run-off.]
(viii) Valuation:	Market price of nitrogen, phosphorus and potash can be used to value the nutrients retained by the soil (Mekuria et al. 2011), following replacement cost approach. Soil carbon may be valued at market price of Emission Reduction Certificate (carbon credit) or replacement cost of compost material or manures.

(B) **Ecosystem service:**	**Climate regulation/ regulating service (Annual average carbon sequestration rate)**
(i) Monitoring indicator:	Quantity of carbon retained/ sequestered by soil and vegetation
(ii) Rationale:	Soils and vegetation can store carbon and soil; and water conservation measures strengthen soil and vegetation condition and, thus, increase this regulating service.
(iii) Indicator:	Amount of carbon stored annually (Mg ha^{-1} year^{-1})
(iv) Data requirements:	Carbon stock in soil, vegetation biomass, Certified Emission Reduction (CER) price $SOC = (\%C \times BD \times Depth \times 10^4)/100$ where, C is percent organic carbon, BD (Mg kg^{-1}) is bulk density; Depth (soil depth in m)
(v) Calculation of the indicator:	Difference in carbon stock after catchment treatment over before treatment or control condition
(vi) Spatial scale:	Specify the area (plots/field, sub-catchment, basin; terraced cropland, agroforestry system-area, grassland, fallow land and where SWC measures were imposed)
(vii) Methodology:	Using appropriate sampling method and laboratory analysis, soil survey may be conducted to determine soil carbon content. In vegetation, the stock includes five carbon pools: above-ground biomass, below-ground biomass, litter, dead wood, and soils (IPCC 2006). Above-ground biomass consists of the living biomass material above the soil. This is multiplied with root-shoot ratio of species to estimate below ground biomass. Using IPCC factor (0.5), carbon content in vegetation may be estimated. $C_{av} = \sum_{i=1}^{n} C_i A_i / A$ C_{av} = Average carbon sequestered in the watershed (Mg ha^{-1}), C_i = average carbon sequestered in the ith land use in the watershed (tha^{-1}), A_i = Area under i^{th} land use in the watershed (ha), A = Area of the watershed [Tree harvested for fuel wood in the watershed may be excluded for this estimation].

(continued)

Table 3.3 (continued)

(viii) Valuation:	Carbon may be converted into CO_2 equivalent by using standard factor and may be multiplied with the price of certified emission reduction (carbon credit) (Mekuria et al. 2011).
(C) **Ecosystem service**	**Soil health maintenance**
(i) Ecosystem function affected by integrated water management and soil water conservation practices	Water resource regulation (*in-situ* moisture), Nutrient regulation (regulating service)
(ii) Monitoring indicator	Soil health (physical, chemical, biological properties)
(iii) Rationale	Resource conservation, land management changes and vegetation improvement due to watershed management interventions affect the soils in the watershed. The management interventions lead to change in nutrients (N, P, K, carbon) contents, soil fertility build up, change in soil physical structure, infiltration rate which affects the soil quality. Soil organic matter directly impacts water infiltration rates, soil aggregate stability and soil structure. One of the by- products of healthy soil is maintenance of soil nutrients as soil organic matter is also a significant source of nutrients.
(iv) Indicator	Soil nutrients (Mg ha^{-1})
(v) Data requirements	Soil nutrients (N, P) content in different land uses; area under different land uses; nitrogen and phosphorus content in soil; watershed area: market price of nitrogen and phosphorus.
(vi) Calculation of the indicator	Difference in nutrient (N, P) content after watershed management intervention over before watershed intervention or control condition.
(vii) Spatial scale	Specify the area (plots/field, sub-catchment, basin, terraced cropland, agroforestry system-area, grassland, fallow land and where SWC measures were imposed)
(viii) Methodology	Assess the total soil nutrient (N, P) content under each land uses before and after the watershed interventions. $NC_{av} = \frac{\sum_{i=1}^{n} NC_i A_i}{A}$ NC_{av}= Average nutrient (N, P) content in the watershed (Mg ha^{-1}), NC_i = nutrient (N, P) content in the ith land use in the watershed (Mg ha^{-1}), A_i = Area under i^{th} land use in the watershed (ha), A = Area of the watershed
(ix) Valuation	The nitrogen and phosphorus content (proxy for soil health) may be valued in terms of the replacement cost of N and P at market prices.

(continued)

Table 3.3 (continued)

(D) **Ecosystem service**	**Supporting service**
(i) Ecosystem function affected by IWM/ SWC practices:	Soil regeneration, In-situ water augmentation
(ii) Monitoring indicator:	Conservation of topsoil layer, supporting soil biota, which is quite productive for supporting vegetation and, thus, production.
(iii) Rationale	Earthworms are the most important component of the soil biota (Breemen and Buurman 2002; Butt 2008) which provides soil formation and mineralization functions. Earthworms also help in the maintenance of soil structure and fertility (Edwards 2004). Their activities bring up sub-surface soil (between 10 and 500 tonnes $ha^{-1}yr^{-1}$), providing nutrients in the plant root zone and aiding the formation of approximately 1 Mg $ha^{-1}year^{-1}$ of topsoil (Pimentel et al. 1995). Soil micro-organisms and invertebrates help break down of organic matter in soil (Brady and Weil 2004; Bashan and de Bashan 2010). This releases organically bound nutrients such as nitrogen for use by plants (Edwards and Arancon 2004). Water and soil conservation provides conducive environment for support and growth of earthworm which helps support the process of soil formation (Mandal et al. 2010).
(iv) Indicator:	Topsoil regenerated and mineralized.
(v) Unit of measurement:	Mg $ha^{-1}year^{-1}$
(vi) Data requirements:	Soil earthworm count.
(vii) Calculation of the indicator:	Difference in soil formation and mineralization (based on soil loss tolerance limit values (Mandal et al. 2006, 2010; Mandal and Sharda 2011) multiplied with value of topsoil before and after the watershed programme or control.
(viii) Spatial scale	Specify the area (plots/field, sub-catchment, basin, terraced cropland, agroforestry system-area, grassland, fallow land and where SWC measures were imposed)
(ix) Methodology:	Soil sample for earthworm count per unit area can be projected for earthworm biomass per hectare. Contribution of earthworms in soil formation can be calculated based on the assumptions that the mean biomass of an earthworm is 0.2g (Fraser et al. 1996) and one tonne of earthworms forms 1000 kg of soil $ha^{-1}year^{-1}$ (Pimentel et al. 1995). This may be projected to assess quantum of soil formation over a given period due to earthworm in the conserved soil. Mineralization rate of organic matter by soil micro-organisms and invertebrates (Brady and Weil 2004) can be done by assessing organic matter (OM) from weight of soil (obtained from bulk density at 10 cm depth) and soil nitrogen from soil testing. Ratio of OM to nitrogen may be taken 20:1 (Brady 1990) and total mineralized nitrogen may be estimated.

(continued)

Table 3.3 (continued)

	$N_{min} = n \times b \times v \times k \times 10^{-3} kg$ N_{min} = amount of N mineralized, n = total amount of N (%) in soil, b = bulk density of soil (g cm^{-1}), v = volume of soil (cm^3), k = percentage of mineralization (%)
(x) Valuation	The quantum of soil (t $ha^{-1}year^{-1}$) could be multiplied with the value of topsoil (market price) by following the method given by Sandhu et al. 2008. The amount of nitrogen mineralized could be multiplied with market price of nitrogen (INR Rs. Per kg) (Sandhu et al. 2008). Benefit transfer method can be used with values from tropical region.

SLMPs (control). Then the value of support service was assigned by following replacement cost technique, i.e. assigning value of 1 Mg of soil as computed by Mandal et al. (2021) and the saving of soil nutrients (N, P, K and S) in the retained sediment were included. Market price of nitrogen, phosphorus and potassium were taken to compute the value of topsoil (Table 3.4). The quantum of soil loss (Mg ha^{-1} $year^{-1}$) was then multiplied with the value of topsoil following market price method (Sandhu et al. 2008).

Resource conservation, land management changes and vegetation improvement due to SLMP interventions affect the soil in the treated area. The management interventions lead to changes in soil fertility build-up, change in soil physical structure, infiltration rate which affects the soil quality. Soil organic matter directly impacts water infiltration rates, soil aggregate stability and soil structure. One of the by-products of healthy soil is maintenance of soil organic matter which is also a significant source of nutrients. Therefore, the regulating services were computed based on soil organic carbon build-up and soil water retention and supply capacity. Soil carbon was valued at market price of 'Emission Reduction Certificate' (carbon credit) or replacement cost of FYM, wherever applicable. In case of water supply through irrigation price was taken into consideration following the 'replacement cost approach'. Reducing run-off encourages water intake to store more water in soil for plant use especially during the period of low rainfall. Reduction in surface run-off of different SLMPs compared to without SLMP was taken into consideration assuming that amount of water enters into the soil and behaves as irrigation water. Wherever, reductions in run-off values are not available, the retention of water by soil organic carbon was considered to compute the cost of irrigation water. Mandal et al. (2021) computed the value of organic carbon that 1 tonne contains 841.37 l m^3 of irrigation water in one hectare area. To meet the required quantity of available water, irrigation needs to be supplied and it is assumed that irrigation is given through electric pump from groundwater. Total cost of irrigation was computed by taking the component costs of electricity used, water cost of the respective state pump cost (assuming the life of pump is 15 years) and maintenance cost as suggested in Indian literatures.

Table 3.4 The price of various fertilizers used in replacement cost approach for computing soil value

Component/ Nutrients	Fertilizer	% Nutrients in Fertilizer	Cost USD/kg (INR.Rs/ kg)	Reference
Nitrogen	Urea	46% N	0.18 (11.56)	Fertiliser statistics 2016–2017, The fertiliser Association of India, New Delhi
Phosphorus	DAP	18% N & 46% P	0.67 (43.27)	Fertiliser statistics 2016–2017, The fertiliser Association of India, New Delhi
Potassium	MOP	60% K_2O	0.29 (18.33)	Fertiliser statistics 2016–2017, The fertiliser Association of India, New Delhi
Sulphur	Zn sulphate monohydrate	15% S	1.36 (87)	https://www.iffcobazar.in/en/product/zinc-sulphate-monohydrate-33-zinc#zinc-fertiliser-5-kg

Note: 1 US$ = 64.2 INS. Rs

$$\text{Total cost} = \text{Electricity cost} + \text{Water cost for irrigation} + \text{Pump cost} \\ + \text{Maintenance cost}$$

In few cases, the SOC values were not available; however, we computed the above-ground and below-ground biomass from reported yield and then estimated the contribution of root biomass to SOC.

3.3 Case Studies in Hill and Mountain Ecosystems of India

3.3.1 Western Himalayan Region

The north-western Himalayas with an area of 33.14 million ha covers nearly 10.07 per cent of the total geographical area (329 M ha) of the country. With varied physiography and climate, the region is endowed with unique soil, water and biotic resources. The region consists of major parts of three states, namely Jammu and Kashmir, Himachal Pradesh and Uttarakhand. The landscape is by and large undulating and slopping except for lower parts in *Bhabar, Tarai* and Valley regions. The climate varies from hot and sub-humid tropical in the southern low tracts to temperate, cold alpines in the northern high mountains. The average yearly rainfall ranged between 350 mm and 3000 mm in different parts of the region; however, the rainfall is extremely low in Leh (92.6 mm). Soils on the hilly and side slopes had a relatively shallow depth than soils of the valley region. Dominating soil type is Entisols (47.3% area) followed by Inceptisols (18.5%), Mollisols (0.6%) and Alfisols (0.4%). Greater Himalayas and Shivalik regions have the highest concentration of organic carbon. The higher organic matter content in high altitudinal regions is due to thick forest vegetation, high rainfall and low decomposition rate. Soils are slightly acidic to moderately alkaline, and soil depth ranges between 25 and 100 cm. Analysis of soil erosion data with different categories and corresponding 'Soil Loss Tolerance Limits (T)-values' revealed that only about 20% area falls under the severe and extremely severe erosion class, whereas about 22.5% area falls between very low to moderate erosion class (Sharda and Mandal 2018). This clearly reflects that soil erosion by water is a major problem in one-third area of the region with soil erosion rate of more than10.0 Mg ha^{-1} $year^{-1}$. Analysis of tolerance limits data revealed that the soil loss tolerance limit values varied between 2.5 and 12.5 Mg ha^{-1} $year^{-1}$ depending on soil depth and soil resistivity to erosion. Only about 1.17% of total geographical area can tolerate a soil loss of less than 2.5 Mg ha^{-1} $year^{-1}$ while about 3.89% area has T-value of 12.5 Mg ha^{-1} $year^{-1}$. Based on the analysis, it is concluded that only about 0.85% of total geographical area of the region requires different degrees of erosion management while 30.42% of total geographical area falls under no treatment category due to higher values of permissible erosion limits (Mandal and Sharda 2013). However, about 13.69% area falls in priority class 1 and 2. The areas where the present erosion rate is exceeding the critical limits require to be treated on priority basis through appropriate SWC so that the erosion rate is reduced to the permissible rates. The acute problem of soil erosion in Shiwalik and high Himalayas is quite evident in large-scale terrain deformation. Increased soil erosion affects the soil productivity and causes many off-site damages in terms of rapid siltation of multipurpose reservoirs and lakes. Therefore, these areas are relatively more sensitive and call for minimal soil disturbance.

3.3.2 Eastern Himalayan Region

This region of India encompasses northern hilly parts of the country with an area of 25.38 million ha, representing 7.71% of the total geographical area of the country. This agroclimatic region is characterized by warm summer and cool winter. The annual precipitation varies from 2000 to 3000 mm. Forest soils are rich in organic matter with moderate to low base status. High rainfall allows intense leaching which results poor base status of the soils. In general, Jhum cultivation is the traditional farming system in about 12% total geographical area of the whole region. It is practised with mixed cropping on the steep slopes under *rainfed* condition at an interval of 3–4 years causing severe erosion. In the hilly areas, vegetables, tea, pineapple, citrus, apple, peer, peach and banana are grown on terraces. The natural vegetation comprises of subtropical conifers along with wet evergreen and subalpine forest. Steeply sloping landforms cause serious run-off resulting in severe erosion. Deforestation in shifting cultivation areas also leads to severe soil degradation problem. Hill terraces are also used for plantation crops, such as tea, coffee, rubber and horticultural crops, like oranges, pineapple, etc. The continuous land degradation would influence adversely the livelihood and ecological steadiness not only in highlands, but also in lowlands.

Analysis of soil erosion data revealed that soil erosion rates varied enormously across the region, ranging from less than 5 Mg ha^{-1} $year^{-1}$ to more than 40 Mg ha^{-1} $year^{-1}$. About 29.82% area in the region falls under the category of very severe erosion, whereas the percentage of very low, low, moderate and severe erosion classes is 4.47, 21.16, 16.79 and 12.97%, respectively (Mandal and Sharda 2013). The results further revealed that due to strong resistance capacity, T-values were consistent at around 7.5 and 12.5 Mg ha^{-1} $year^{-1}$ except for about 0.26% area with T-value of 5.0 Mg ha^{-1} $year^{-1}$. The areas of major concern are soils on ridges and side slopes which can tolerate only a low soil loss up to 7.5 Mg ha^{-1} $year^{-1}$. Other critical areas are West Tripura, Dhalai and Darjeeling hills which can withstand a soil erosion rate of only 5.0 Mg ha^{-1} $year^{-1}$. Therefore, the above areas are most susceptible and need due attention during crop cultivation with monoculture. Based on the analysis, it is projected that about 58.94% of total geographical area (TGA) of the region requires different degrees of erosion management and 28.38% of TGA falls under no treatment category due to higher values of permissible erosion. Around 42.79% area of the region falls under severe and very severe erosion categories, 65.09% has high priority for conservation treatment as the soil loss tolerance limit is less than 10.0 Mg ha^{-1} $year^{-1}$. However, the assessment based on comparison between prevailing and permissible rate of erosion indicates that nearly 29.70% area of the region is very critical as classified up to priority class 2. In general, the areas where the current soil erosion rate exceeds the tolerance limits need to be addressed on priority through suitable SWC measures. Successful planning and implementation of soil conservation programmes are essential as farming has been extended to slopping-marginal lands through widespread deforestation ensuing in accelerated soil erosion and low crop yield. Indiscriminate deforestation and practice of *Jhum* cultivation led to accelerated erosion for which proper conservation measure needs to be established especially on very steep slopes.

3.3.3 Cost of Ecosystem Services (ES) in Hill and Mountain Ecosystem

In this region, total 43 studies were considered to compute ecosystem services (ESs) of various SLMPs. The analysis revealed that compared to zero tillage (N = 6), the conventional tillage + mulch ($N = 6$) had higher ESs. Wheat crop had additional ecosystem benefit of USD 198 (INR. Rs. 14,540) to USD 331 (INR.Rs. 24,307); however, the additional benefits in maize with only mulch treatment varied between USD 36 (INR.Rs. 2636) and USD 108 (INR.Rs. 7931) (Table 3.5). *Toria,* an oilseed crop had additional benefits of USD 26 (INR. 1927) to USD 64 (INR.Rs. 4675) in conventional till + mulch condition. Combination of tillage and mulch had the highest values with respect to regulating services particularly for water regulation which varied between USD 25 (INR.Rs.1807) to USD 44.5 (INR. 3272) in wheat and USD 16 (INR.Rs. 1137) to USD 29.5 (INR.Rs. 2166) in maize crop. In hill and mountain areas, soil and water conservation should emphasize for mitigating soil erosion through cost-effective methods, which reduce overland flow and improve

Table 3.5 Cost of ecosystem services (ESs) of various soil land management practices (SLMPs) in hill and mountain region of India

<table>
<tr><td colspan="7">Himalayan Region</td></tr>
<tr><td colspan="7">1. SWC measures (agronomic or vegetative measures)</td></tr>
<tr><td colspan="2" rowspan="2">• Change over control
(N = 17)</td><td rowspan="2">Provisioning services (yield increase) USD/ha (INR/ha)</td><td rowspan="2">Supportive services (soil loss reduction), USD/ha (INR/ha)</td><td colspan="2">Regulating services</td><td rowspan="2">Total services, USD/ha (INR/ha)</td></tr>
<tr><td>Water regulation, USD/ha (INR / ha)</td><td>Organic carbon storage, USD/ha (Rs/ha)</td></tr>
<tr><td rowspan="10">(a) Agronomical measures
(N = 17)</td><td rowspan="2">Wheat (tillage+mulch) vs Zero tillage
(N = 3)</td><td rowspan="2">166 (12318.5)
to
276 (20299.5)</td><td rowspan="2">5.6 (413.85) to 10 (735.54)</td><td>15.5 (1136.85) to 29.5 (2166.45)</td><td>9.13 (671.03) to 15 (1105.79)</td><td rowspan="2">198 (14540.23)
to
331 (24307.28)</td></tr>
<tr><td colspan="2">25 (1807.88) to 44.5 (3272.24)</td></tr>
<tr><td>Maize (tillage+mulch) vs Zero tillage
(N = 3)</td><td>–</td><td>5.6 (413.85) to 10 (735.54)</td><td>15.5 (1136.85) to 29.5 (2166.45)</td><td>–</td><td>21 (1550.70) to 39.5 (2901.99)</td></tr>
<tr><td rowspan="2">Maize (mulch) vs without mulch
(N = 3)</td><td rowspan="2">29 (2137.5) to 89.2 (6555.0)</td><td rowspan="2">1.15 (85.33) to 5.8 (426.65)</td><td>3.4 (249.89) to 6 (448.08)</td><td>2.2 (163.58) to 6.8 (501.64)</td><td rowspan="2">36 (2636.3) to 108 (7931.37)</td></tr>
<tr><td colspan="2">5.6 (413.47) to 13 (949.72)</td></tr>
<tr><td rowspan="2">Maize
(tillage+crop residue) vs Conventional tillage
(N = 4)</td><td rowspan="2">11.6 (855.0) to 38.77 (2850.0)</td><td rowspan="2">2.3 (170.66) to 4.8 (358.39)</td><td>2 (150.20) to 4 (300.30)</td><td>0.9 (65.43) to 3 (218.10)</td><td rowspan="2">16.8 (1241.29) to 50.7 (3726.79)</td></tr>
<tr><td colspan="2">3 (215.63) to 7 (518.4)</td></tr>
<tr><td rowspan="2">Toria
(tillage+crop residue) vs Conventional tillage
(N = 4)</td><td rowspan="2">21.2 (1560.0) to 53 (3900.0)</td><td rowspan="2">2.3 (170.66)
to
4.8 (358.39)</td><td>2 (150.20) to 4 (300.30)</td><td>0.6 (46.53) to 1.6 (116.32)</td><td rowspan="2">26 (1927.39) to 63 (4675.01)</td></tr>
<tr><td colspan="2">2.7 (196.73)
To
5.6 (416.62)</td></tr>
</table>

2. Agroforestry or land use measures						
• Change over control ($N = 26$)		Provisioning services (yield increase) USD/ha (INR/ha)	Supportive services (soil loss reduction), USD/ha (INR/ha)	Regulating services		Total services, USD/ha (INR/ha)
				Water recharge, USD/ha (INR/ha)	Organic carbon storage, USD/ha (INR/ha)	
Crop cover management	Maize through legume intercropping vs. maize as control ($N = 5$)	90 (6626.25) to 210 (15461.25)	2.2 (164.69) to 6 (441.16)	1.3 (95.54) to 10.2 (750.68)	20 (1463.12) to 231 (16984.86)	114 (8349.6) to 458 (33637.95)
				21 (1558.66) to 241 (17735.54)		
Tea plantation vs. open pine forest ($N = 3$)		–	2.7 (200.81)	14.5 (1062.58)	3 (220.20)	20.2 (1483.59)
				17.5 (1282.78)		
Rainfed agriculture vs. open pine forest ($N = 3$)		–	3.54 (260.54)	18.5 (1360.24)	3.5 (257.65)	25.5 (1878.43)
				1617.89		
Grazing land vs. open pine Forest ($N = 3$)		–	2.45 (180.33)	8.34 (612.90)	3 (214.31)	13.70 (1007.54)
				11.3 (827.21)		
Forest vs. agriculture ($N = 2$)	14 (1000)	–	–	–	368 (27046.74)	
Forest vs. agriculture ($N = 2$)	19 (1400)	–	–	–	–	
Forest vs. agriculture ($N = 2$)	22 (1600)	–	–	–	258 (18956.89)	
Wasteland vs. agriculture (N = 2)	**14 (1000)**	–	–	–	–	

(continued)

Table 3.5 (continued)

Himalayan Region						
Wasteland vs. agriculture ($N = 2$)	19 (1400)	–	–	–	–	
Wasteland vs. agriculture ($N = 2$)	22 (1600)	–	–	–	–	

infiltration of excess water. Even in such areas, the excess water may derive some benefits for supplementary irrigation during the post-rainy season or for domestic and livestock use (Joshi et al. 2005).

Land use land cover change through legume intercropping in maize had the highest additional ESs to the tune of USD 457 (INR. 33,637) compared to without intercropping. Interestingly, here the contribution of regulating service was more (to the tune of USD 241; INR.Rs. 17,735) than the provisioning service (USD 210; INR.Rs. 15,461) (Table 3.5). In this region, any agronomic measure which has the capacity to protect topsoil is very effective. Increasing the SOC pools of degraded soils would increase crop yields by influencing available water retention. There exists a strong relationship between the SOC pools and the plant available water retention and the ability of soils to withstand drought (Hudson 1994; Emerson 1995). Moreover, soil erosion unlocks the SOC and thereby increases emissions of CO_2, exacerbating global warming (Lugato et al. 2018; Mandal et al. 2020). Therefore, SLMPs help in carbon sequestration also. In other land use systems, such as tea plantation, rainfed croplands and grazing lands, the cost of provisioning service had not been computed due to non-availability of data. However, *rainfed* cropland showed higher values of regulating services (USD23.4; INR. 1718) compared to tea plantation (USD 17.4; INR.Rs. 1282) and grazing lands (USD 11; INR. 827). Similarly, for shifting cultivation areas in north-eastern Himalayan region (NEHR), the additional benefits of regulating service varied between USD 32 (INR. 2358) and USD 62.2 (INR. 4572). This computation was made for only water regulating as the soil carbon values were not available; the computation of carbon regulating service could not be made. Land restoration in shifting cultivation areas significantly reduces the soil erosion and surface run-off, therefore there are high possibilities that the soil supporting service through retention of soil and water regulating through increasing infiltration would greatly improve. The additional ecosystem benefits through soil retention were USD 34 (INR. 2519) in grasslands and USD 32 (INR.2358) in forest lands against *Jhum* lands (Table 3.6) (Mandal and Sharda 2013).

Table 3.6 Land restoration in shifting cultivation area in north-east Himalayan region

SLMPs		Provisioning Services USD/ha (INR/ha)	Supporting Services USD/ha (INR/ha)		Total Services, USD/ha (INR/ha)
Land under forest vs. Jhum	–	34 (2499.32)	33 (2358.20)	–	67 (4857.52)
Settled agriculture vs. Jhum	–	12 (881.03)	54 (3976.70)	–	67 (4857.73)
Natural grasses vs. Jhum	–	34.3 (2519.79)	62.2 (4572.10)	–	96.5 (7091.89)

3.4 Valuation of Participatory Watershed Management with Case Studies

A watershed is a mixed production system where plants (crops, horticulture, grasses and forages) and animals (livestock, poultry and fisheries) interact within a biogeochemical human-managed ecosystem (Altieri 2002). It is not always possible to protect the entire watershed because of high anthropogenic pressure. General recommendation is that high erosion risk areas with higher slopes should be converted to forest and low erosion risk areas may be transformed with agroforestry or terraced croplands for better soil and water management and biodiversity conservation. An 'Integrated Natural Sequence Farming' method has been developed in north-eastern Himalayan region of India where the top 1.5 ha area was kept under protected forest having 800 m^3 capacity of water storage in the middle-reaches with provision of well-planned silt retention ditches and run-off water course. In that watershed, lower parts were used for paddy-cum-fish culture for about 1.9 ha areas. Split bamboo channels were used for water conveyance. That type of nature-based solutions has a lot of benefits in terms of ecosystem services. Although agroecosystems were normally managed in isolation within a region, however, that hill agroecosystem has numerous consequences in the neighbouring systems.

The following points are very important in Integrated 'Natural Sequence Farming':

1. Recycling of biomass and balancing of nutrients flow.
2. Securing favourable soil conditions for plant growth through increasing soil organic carbon content and soil biotic activities.
3. Minimizing losses of solar radiation, air, water and nutrients (soil surface management, microclimate management and water harvesting).
4. Enhancing biodiversity (SALT for hilly areas, sequential farming; agroforestry, hedge row cultivation).
5. Enhancing beneficial biological interaction and synergisms among agrobiodiversity components.

3.4.1 A Case Study on the Pricing of 'Participatory Watershed Management' in Hill and Mountain Region in Himalayas, India

The integrated watershed management projects are more appropriate for balanced development than sectoral or commodity approach (Joshi et al. 2005; Dhyani et al. 2006; Raizada et al. 2008; NRAA 2011). An attempt was made to quantify all the four ESs (provisioning, regulatory, supportive and cultural) at one of the oldest micro-watersheds (370 ha) management project at Fakot in north-western Himalayas (NWH), India to demonstrate the efficiency of participatory integrated watershed approach for simultaneous improvement of all ESs over the years (Dhyani et al. 2021).

Table 3.7 Watershed interventions implemented during two periods of watershed project

Interventions	Unit	1975–1986 (IP)	1987–2013 (PIP)
Cement lined tank (each having 20 to 60 cum capacity)	Number	7.0	6.0
Cement lined channel	Metre	2316.0	2489.0
Irrigated terrace renovation	Hectare	7.9	0
Conversion of rainfed terraces into irrigated	Hectare	12.8	5.5
Renovation of rainfed terraces	Hectare	20.0	12.5
Horticultural plantation	Hectare	21.2	9.1
Grassland developed	Hectare	1.5	72.0
Fuel-fodder plantation	Hectare	22.4	3.6[a]
Gully plugs	Number	66	138

[a]Excludes the naturally regenerated tree area. Crop demonstrations >5000 number (total)

All interventions pertaining to crop, horticulture and pasture development were monitored through simple random sampling technique by adopting with and without project approach and evaluated through budgeting method till 2013 (Table 3.7). Land use budgeting along with crop varieties were suitably integrated for estimating agricultural production (food, fruit and fodder). Complete enumeration of farmers was performed at appropriate time intervals to estimate income, outmigration and livestock production. Rainfall data were monitored with Indian Meteorological Department (IMD) standard rainfall recording equipment. Run-off and soil loss data were monitored at the watershed outlet through a run-off gauging station of 'United States Department of Agriculture' (USDA) standard. Concurrent evaluation results were utilized to modify the project interventions with legitimacy, and most adoptable efficient technologies were upscaled. After 1986, project interventions on livestock improvement, drainage line treatment, water harvesting structures, landslide control measures and crop demonstrations of latest varieties and integrated pest management were undertaken by convergence of other organizations/programmes like 'Integrated Rural Development Programme', 'Border Road Organization', state agricultural universities and concerned state line departments. The technical backstopping and monitoring of watershed responses were continued till 2013. The period (1987–2013) was also utilized to understand the dynamic change-chain process in real farm situation. All primary and secondary data available from project records and reports were utilized for quantification of ESs from Fakot watershed (Dhyani et al. 2021).

The indicators estimated under each group of ESs from the project are presented in Table 3.8.

Table 3.8 Ecosystem services and indicators estimated

S. No.	ESS group	Service	Indicator(s)
i.	Provisioning services	Food production	Rice Equivalent total crop Production (REP) $REP = \sum (y_i * p_i)/p_r$, where y_i is production of ith crop, p_i is price of ith crop, p_r is price of rice. Similarly for horticultural crops, Mango Equivalent Production (MEP) was considered.
		Fodder and fuel availability	Dependency on reserve forest for fuel and fodder (FD) $FD = \frac{\text{Fodder/fuel collected from forest}}{\text{Total fodder/fuel demand}} * 100$
ii.	Regulating services	Water regulation	Conserved rainwater, reduction in run-off peak, drinking water quality based on comparison of 'before and after' watershed interventions.
		Erosion control	Reduction in soil loss based on comparison of 'before and after' watershed interventions.
iii.	Cultural services	Social relations	Community organizations, group activity, outmigration status and women empowerment.
		Educational value	Change in attitude and visitors visited
iv.	Supportive services	Nutrient cycling	Availability of N, P_2O_5 and K_2O, availability of micronutrients
		Biodiversity	Vegetation diversity index, crop diversification index

3.4.2 Observation Recorded and Data Collected of Pricing of ES of Integrated Watershed Management Project

Benchmark data on soil, vegetation, land-use pattern with productivity, water resources, animal husbandry, land degradation and socio-economic parameters were collected through traditional surveying techniques required for each aspect of integrated watershed planning in 1974–1975. Simple budgeting technique was employed to derive average watershed situation with spatial distribution and has been named as 'Before Project' period (BP). Field level data on land use pattern, adoption of improved natural resource management and production technologies, production from annual and horticultural crops, livestock products, fuel and fodder production, rainfall, run-off and soil loss, farm income, group activities, outmigration, and water availability with quality were monitored regularly during 1976 to 2013 by adopting stratified simple random sampling technique with a methodology as prescribed by respective indicator discipline. Data collected during 1976 to 2013 were grouped into two time periods. Field data on asset value, role of women in decision-making and children school enrolment were collected during the year 1974–1975 (BP), 1986–1987 (Project initiation/implementation period, PIP) and 2012–2013 (last year of PIP). Data on soil fertility and organic carbon were collected at two points of time, i.e. during project period (PP) and on 2007–2008 as they were not supposed to change in a short span of time. Asset value index, percentage of children of 5 to 16 years going to school, soil fertility status and carbon sequestration were recorded during the year of data collection only, based on 'before and after approach'. Therefore, they are point estimate only. Data for estimating vegetation diversity or richness indices were estimated by adopting with and without project approach and data were collected from Fakot (treated) watershed and adjacent Agar (non-treated) watershed in 2009 only. Thus, results are based on a mix set of data over 38 years and transdisciplinary methodologies of ESs estimation.

Vegetation data collected during PP were found to be of little use for estimating vegetation diversity and richness indices by an independent evaluation agency. Therefore, evaluation agency collected required data in 2009 by adopting with and without project approach (GOI 2011). Data were analysed by adopting standard analytical procedures applicable to each indicator discipline. Therefore, the data analysis methodology followed in the study is termed as 'Toolkit Method'.

Consequently, natural regeneration in the reserve forest, as well as wasteland within the watershed, took place and it was able to deliver intended ESs in due course of time (GOI 2011). More water availability after rainy season (July–September) and cultivation of short duration crops increased gross irrigated area over PP by more than 100% and 300% during BP and PIP, respectively (Dhyani et al. 2021).

The provisioning service improved significantly after project period (Table 3.9).

The change in indicator(s) value can be computed from comparative data recorded during before and after project period. The change in improvement or loss of ESs can then be computed by multiplying it with assigned monetary value. An example for computation of travel cost and additional fodder grown through successful intervention of watershed (SWC) is given below. Similarly, the value can be assigned for all other indicators of respective ES (Table 3.10).

Table 3.9 Level of provisioning service indicators during three periods of project

S. No.	Indicator	Measurement unit	Before project (1975–1986)[a]	After project (1987–2013)[a]
i.	Rice equivalent total food production	Mg	398.6	714.2
ii.	Mango equivalent total fruit production	Mg	6.4	232.1
iii.	Total milk production	Kl	184.8	473.2
iv.	Dependency on reserve forest for fodder	%	46	5
v.	Dependency on reserve forest for fuel	%	70	20
vi.	Average gross irrigated area	ha	34.8	69.8

[a]Average value of the period

Table 3.10. Price of additional fodder production (Change in productivity and travel cost approach)

Particulars	Before project	After project
Fodder requirement q/year	33,163	52,800
Brought from forest q/year	19,898	10,599
Produced from agriculture + community land q/year	13,265	42,241
Additional quantity produced from agriculture + community land = 42,241–13,265 = 28,976 q/year		
Market value of fodder @USD 1.66/q (INR.Rs. 100/q) = 28976x100 = 48,293 (INR.Rs. 2,897,600/)		
Average head load per trip = 0.4 q		
Total no. of trips = 28,976 ÷ 0.4 = 72,440		
Travel time saved per trip = 30 minutes Total time saved = (72440x30)/60 = **36,220 hours.**		
No. of days saved @ 8 hours/day = 36,220 ÷ 8 = 45,275 days		
Price value of time saved (wage rate is USD 2.5/day (INR.Rs. 150/day) but it was reduced to 50% as opportunity of labour for this work is less) @ USD 1.25/day (INR.Rs. 75/day) = USD. 56,594 (INR. Rs. 3,395,625/–		

q Quintal; 1 USD = 60 INS. Rs

Total soil carbon sequestrated by the watershed before the project was 18911.22 Mg and reached to the level 19524.49 Mg in 2007. Thus, there was an increase of 613.27 Mg soil carbon stock at 0 to 30 cm soil depth of the watershed in 2007 as compared to 1974. Soil organic carbon percentage (SOC%) decreased under uncultivable lands at 0 to 15 cm soil depth and remained unchanged in 15–30 cm soil depth. In cultivated area good soil carbon was built at 0 to 15 cm soil depth and a marginal decrease in the 15 to 30 soil layer was observed. It was also observed that soil carbon build-up was more in valley than upper and middle reaches and in the order of grassland> horticultural > forest land uses (Table 3.11).

The ecosystem benefits due to soil and water conservation intervention in the watershed can be realized by understanding the saving of travel time for fodder and fuel collection, soil retention through erosion control, nutrient build-up and carbon sequestration. Earlier (before the watershed intervention) the women community used to travel to near-by forest areas for the purpose of collecting fuel-woods and fodder. After 27 years, they realized the importance of soil and water conservation intervention in terms of many provisioning and intangible benefits. Although, more detail information and computation methodology need to be established to compute the all other benefits, however, here we present some ecosystem benefits in monetary terms of the Fakot watershed (Table 3.12).

Table 3.11. Valuation (pricing in monetary term) of soil carbon build-up in different land use and land management practices in watershed

Land use / land management practices	Area (ha)	Soil depth (cm)	Carbon sequestration (Mg ha^{-1})	Cost of carbon sequestration USD $; (INR.Rs)
Renovation of irrigated terraces	7.4	0–15 15–30	11.31 7.29	1758; (105481.17)
Conversion of rainfed terraces into irrigated terraces	12.8	0–15 15–30	9.20 9.02	2961; (177675.60)
Renovation of rainfed terraces	20.0	0–15 15–30	7.94 7.63	3954; (237240.09)
Horticultural plantation in wasteland	14.2	0–15 15–30	18.30 15.21	6042; (362520)
Horticultural plantation in rainfed terraces	7.6	0–15 15–30	18.09 9.02	2616; (156968.52)
Fuel-fodder plantation in wasteland	22.4	0–15 15–30	18.97 17.20	10,287; (617256.96)
Forest land	81	0–15 15–30	11.88 11.35	23,892; (1433519.81)
Total				**51,511; (3090662.15)**

Table 3.12. Value of ecosystem services through participatory watershed development at Fakot in western Indian Himalaya

Ecosystem attribute	Valuation approach employed	Value USD; (INR. Rs.)
Additional fodder produced	Travel cost	2264; (135825) yearly
Enhanced fuel availability	Travel cost	590; (35420) yearly
Soil retention	Hedonic	409; (24,542) yearly
Checking of outmigration	Relocation	13,860; (8,31,600) yearly
Soil nutrients build-up	Shadow pricing (replacement cost)	10,495; (6,29,684) whole project period
Carbon sequestration	Shadow pricing (replacement cost)	51,511; (3090662) whole project period
	Grand Total	**79,129; (4747740)**

3.4.3 Ecosystem Disservices

Loss of critically important ecosystem services can be termed as ecosystem disservices. In designing a watershed management programme, the entire ecosystem must be taken into consideration. Poorly designed watershed management programme or soil water conservation or carbon sequestration projects could negatively impact the ecosystems. For example, monoculture enhances soil erosion and deteriorates soil quality. Similarly, watershed projects that measure success in terms of water flow may create incentives to divert water from the irrigation of local crops to downstream water delivery in a drought year, jeopardizing subsistence farmers.

The payments for ecosystem services (PES) on watershed basis currently exist in Costa Rica, Ecuador, Bolivia, India, South Africa, Mexico and the United States. In most of those cases, maximizing ecosystem services through payment systems has led to poverty alleviation (Asquith et al. 2007; Agarwal and Ferraro 2007). This clearly established that the potential trade-offs between poverty reduction and ecosystem should be harnessed. Practitioners and policymakers around the world have already proven that they can design and implement PES programmes that optimize the benefits. In most of the cases PES initiatives are (by definition) voluntary because they involve transfers of wealth (often from wealthier urban areas to poorer rural areas), and because they can empower the poor by recognizing them as valued service deliverers. The PES schemes are more likely to have pro-poor impacts than most other environmental management interventions (Bruijnzeel and Von Noordwijk 2007).

3.5 Future Road Map

The last few years have bestowed us with improved and clear data on land use pattern carbon stocks along with information about flow within the watershed which makes the study of effects of various management systems on soil ecosystem services extremely fascinating. Besides this, understanding how climate change affects the supply and economic values of services at a watershed scale is also pertinent to grasp the biophysical and economical aspects of soil. Among the handful of successful projects, one that especially stands out while covering this aspect of physical and economical flows of soil ecosystem services is the SoilTrEC project (Banwart et al. 2012). This chapter proposes a soil ES framework that focuses on promoting the sustainable management of soils. The highlight of the framework is the holistic approach towards elucidating the economic benefits of different soil ES and estimating the cost of soil degradation. It also considers the associated loss of soil ES. Availability of such information makes it easier for land managers and policymakers, to ascertain any intrinsic dangers associated with relying solely on economic values of soil ES, as well as gives them a better understanding of any possible trade-offs in land-use management. There is also the issue of any negative consequences, in the long run, rising due to the danger of pivoting solely on provisioning services that are usually the highest yielding service in economic value. These consequences may be related to soil biodiversity accompanying other soil functions and services and can lead to the mismanagement of the soil resource which will result in incurring consequential economic costs in the future. Furthermore, young soil containing underdeveloped soil functions and services is more likely to be perceived as low value and therefore, deemed unimportant. Hence, it must be stressed that the economic value of soil ES not only delivers associated important information but also should be presented contextually with other metrics that are inherent for sustainable land management (Jónsson and Davíðsdóttir 2016).

Evaluating such programmes using the developed indicators will increase transparency and accountability to the public, as well as instil greater confidence in the implementing institutions. The indicators would also allow for a more scientific and systematic evaluation of various watershed development initiatives carried out by different developmental agencies in terms of performance and impact across watersheds within the state, region and country. The Government of India may make the use of these indicators a requirement in all ongoing watershed development projects as part of the Common Guidelines for Watershed Development Projects. To accurately measure the efficacy of watershed development programmes and their inter-comparisons, the indicators must be accepted and evaluated on a larger scale, representing varied climatic, physiographic, edaphic and socio-economic circumstances in the country.

3.6 Messages

Ecosystems and their processes are critical to human well-being, and existence. Ecosystem services can be valued economically to help the environmental policy and management at the local, regional and global levels. The use of economic pricing of ecosystem services in decision-making has been hampered by the significant uncertainties associated with monetary estimations. The uncertainties emerge from the many different methodologies, data inconsistencies, and underlying assumptions employed to capture complex ecosystem processes and functions, as well as the transfer of unit values from a few well-studied regions to data-poor places. Therefore, consistent well-accepted site-specific approaches are required. For example, studies in Ontario, Canada, the valuation methodologies and obstacles are illustrated, with a focus on the requirement to establish monetary unit values that are founded in local reality.

To increase the use of the idea in decision-making at different stages, increasing soil natural capital, and understanding the dynamic character of soils are the keys to a balanced provision of ecosystem services. This study discusses many methods for quantifying soil ecosystems and ensuring their long-term viability.

The significance of soil in ecosystem functioning is well established. However, only a few studies linked soil attributes to ecosystem services in the 1990s, when ecosystem services research was primarily focused on defining the idea and framework. Soil ecosystem services are influenced by soil usage and management and are dependent on soil features and interactions. Soil degradation is a severe global concern for food security and ecosystem sustainability due to landslides, erosion and a reduction in soil carbon and biodiversity. Soils' contribution to human well-being goes beyond food production, and this may be addressed by incorporating soils into the ecosystem services framework and linking it to the diverse functions it performs. Much research has been done on soil and ecosystem services, but not all of them have looked at the direct relationship between soil qualities and ecosystem services.

The importance of using soil data in all ES modelling research should be promoted in the future. Soil scientists should collaborate with experts from other fields to promote the importance of soils in the delivery of ecosystem services. Considering the United Nations' Sustainable Development Goals, future soil and ecosystem services research should concentrate on soil functions. It should emphasize soil security's multifaceted function in long-term environmental policy and management. Future soil ES studies should be based on the UN's requirement for inter- and transdisciplinary soil function research techniques to meet the UN's sustainable development goals. Furthermore, it is critical to redefine soil realization and communicate in such a way that policymakers recognize the importance of soils to environmental sustainability and human well-being.

With the advancement and interactions of various scientific disciplines, an elite group of policy advocates, planners and the public agreed that the pattern of harnessing more provisioning services from an ecosystem to meet the tangible benefit of a growing population has harmed other types of ESs. Anthropogenic activity that results in land-use change (for any reason) is thought to be a key driver altering the ESs supply change. Braat et al. (2008) presented a provocative visualization of the trade-off between provisioning and other ESs as the land-use intensity increased, indicating that if an ecosystem is managed primarily for enhancing production/productivity of a single group of ES (particularly provisioning services by intensive land use), others are nearly always negatively affected; and it was observed that improper intensive land-use practices for a long period negatively affect others (ESs). As a result, in the twenty-first century, ES-related considerations are factored into the bulk of land-use decision-making processes. Following MEA (2005), researchers have discovered new ways to measure and project the impact of anthropogenic actions and policy decisions on ecosystem structure, process and various services offered for human well-being.

References

Agarwal C, Ferraro P (2007) Draft paper prepared for the Bellagio expert meeting, sponsored by Fundación Natura Bolivia, IIED, CIFOR and the Eco Fund Foundation Ecuador.

Altieri MA (2002) Agroecology: the science of natural resource management for poor farmers in marginal environments. Agric Ecosyst Environ 93(1–3):1–24

Asquith N, et al. (2007) Global experiences with payments for watershed services: major challenges and solutions. Natura Bolivia/IIED/CIFOR. Available at www.naturabolivia.org

Banwart S, Menon M, Bernasconi SM, Bloem J, Blum WE, de Souza DM, Davidsdotir B, Duffy C, Lair GJ, Kram P, Lamacova A (2012) Soil processes and functions across an international network of critical zone observatories: introduction to experimental methods and initial results. Compt Rendus Geosci 344(11–12):758–772

Bashan Y, de Bashan LE (2010) Microbial populations of arid lands and their potential for restoration of deserts. In: Soil biology and agriculture in the tropics. Springer, Berlin, Heidelberg, pp 109–137

Braat LC, ten Brink PE, Klok TC (2008) The cost of policy inaction: the case of not meeting the 2010 biodiversity target. Alterra.

Brady NC (1990) The nature and properties of soils. Macmillan Publishing Company, New york

Brady NC, Weil RR (2004) Elements of the nature and properties of soils. Prentice Hall, New Jersey

Breemen NV, Buurman P (2002) Soil formation. Kluwer Academic Publishers, Dordrecht

Bruijnzeel LA, Von Noordwijk M (2007) Draft prepared for the Bellagio march 2007 expert meeting. "Bellagio tropical land use and hydrology: what do we know and is it enough?" Gathering sponsored by the Rockefeller Foundation.

Butt K (2008) Earthworms in soil restoration: lessons learned from United Kingdom Case studies of land reclamation. Restor Ecol 16(4):637–641. https://doi.org/10.1111/j.1526-100X.2008.00483.x

Dhyani BL, Dogra P, Mandal D, Kumar S (2021) Ecosystem services from participatory integrated micro-watershed development project in Indian north western Himalayas: part-I. Indian Journal of Soil Conservation 48(2):139–148

Dhyani BL, Raizada A, Dogra P (2006) Impact of watershed development and land use dynamics on agricultural productivity and socio-economic status of farmers in Central Himalayas. Ind J Soil Cons 34(2):129–133

EC (2006) Thematic strategy for soil protection. In: Communication from the Commission to the Council, the European Parliament, the European Economic and Social Committee and the Committee of the Regions. European Commission, Brussels. COM 231

Edwards S (2004) Financial openness, sudden stops, and current-account reversals. Am Econ Rev 94(2):59–64

Edwards CA, Arancon NQ (2004) Interactions among organic matter, earthworms, and microorganisms in promoting plant growth. In: Magdoff F, Weil RR (eds) Soil organic matter in sustainable agriculture. CRC Press, Florida, pp 327–376

Emerson WW (1995) Water retention, organic carbon and soil texture. Aust J Soil Res 33:241–251

FAO and ITPS (2015) Status of the World's soil resources (SWSR)–Main report. Food and Agriculture Organization of the United Nations and Intergovernmental Technical Panel on Soils, Rome, Italy

Fraser PM, Williams PH, Haynes RJ (1996) Earthworm species, population size and biomass under different cropping systems across the Canterbury Plains, New Zealand. Appl Soil Ecol 3:49–57

Hudson BD (1994) Soil organic matter and available water capacity. J Soil Water Conserv 48:188–193

Hussein MA (2008) Costs of environmental degradation: an analysis in the Middle East and North Africa region. Manag Environ Qual 19:305–317

India UI. Government of India (2011) Ministry of statistics and Programme implementation central statistics division, New Delhi

IPCC (2006) IPCC guidelines for National Greenhouse gas Inventories, prepared by the National Greenhouse gas Inventories Programme. In: Eggleston HS, Buendia L, Miwa K, Ngara T, Tanabe K (eds) Agriculture, forestry and other land use 4: Hayama. Japan, IGES

Jónsson JÖ, Davíðsdóttir B (2016) Classification and valuation of soil ecosystem services. Agric Syst 145:24–38

Joshi PK, Jha AK, Wani SP, Joshi L, Shiyani RL (2005) Meta analysis to assess impact of watershed program and people's action. In: Comprehensive assessment research report 8. Colombo, International Water Management Institute

Lefroy RD, Bechstedt HD, Rais M (2000) Indicators for sustainable land management based on farmer surveys in Vietnam, Indonesia, and Thailand. Agric Ecosyst Environ 81:137–146

Lugato E, Smith P, Borrelli P, Panagos P, Ballabio C, Orgiazzi A, Fernandez-Ugalde O, Montanarella L, Jones A (2018) Soil erosion is unlikely to drive a future carbon sink in Europe. Science. Advances 4(11):eaau3523

Mandal D, Dadhwal KS, Khola OPS, Dhyani BL (2006) Adjusted T values for conservation planning in Northwest Himalayas of India. J Soil Water Conserv 61(6):391–339

Mandal D, Giri N, Srivastava P (2020) The magnitude of erosion-induced carbon (C) flux and C-sequestration potential of eroded lands in India. Eur J Soil Sci 71:151–168. https://doi.org/10.1111/ejss.12886

Mandal D, Roy T, Kumar G, Yadav D (2021) Loss of soil nutrients and financial prejudice of accelerated soil loss in India. Ind J Fertilizers 17(12):1286–1295

Mandal D, Sharda VN (2011) Assessment of permissible soil loss in India employing a quantitative bio-physical model. Curr Sci 100(3):383–390

Mandal D, Sharda VN (2013) Appraisal of soil erosion risk in the eastern Himalayan region of India for soil conservation planning. Land Degrad Dev 24:430–437

Mandal D, Sharda VN, Tripathi KP (2010) Relative efficacy of two biophysical approaches to assess soil loss tolerance for Doon Valley soils of India. J Soil Water Conserv 65(1):42–49

Mekuria W, Veldkamp E, Tilahun M, Olschewski R (2011) Economic valuation of land restoration: the case of exclosures established on communal grazing lands in Tigray, Ethiopia. Land Degrad Dev 22:334–344

Millennium ecosystem assessment ME (2005) Ecosystems and human Well-being. Island Press, Washington, DC

NRAA (National Rainfed Area Authority) (2011) Common guidelines for watershed development pr ojects, NRAA, planning commission. Govt. of India, New Delhi

Pimentel D, Harvey C, Resosudarmo P, Sinclair K, Kurz D, McNair M, Crist S, Shpritz L, Fitton L, Saffouri R, Blair R (1995) Environmental and economic costs of soil erosion and conservation benefits. Science 267:1117–1125

Raizada A, Dogra P, Dhyani BL (2008) Assessment of a multi-objective decision support system generated land use plan on forest fodder dependency in a Himalayan watershed. Environ Model Softw 23(9):1171–1181

Sandhu H, Wratten S, Cullen R, Case B (2008) The future of farming: the value of ecosystem services in conventional and organic arable land. An Exp Appr Ecol Econ 64:835–848. https://doi.org/10.1016/j.ecolecon.2007.05.007

Saxena KG, Maihkuri RK, Rao KS (2005) Changes in agricultural biodiversity: implications for sustainable livelihood in the Himalaya. J Mountain Sci 2(1):23–31

Sharda VN, Dogra P (2013) Assessment of productivity and monetary losses due to water erosion in rainfed crops across different states of India for prioritization and conservation planning. Agric Res 2(4):382–392

Sharda VN, Mandal D (2018) Prioritization and field validation of erosion risk areas for combating land degradation in North Western Himalayas. Catena 164:71–78

Sharda VN, Ojasvi PR (2016) A revised soil erosion budget for India: role of reservoir sedimentation and land-use protection measures. Earth Surf Process Landf 41(14):2007–2023

Sharma PD (2004) Managing natural resources in the Indian Himalayas. J Ind Soc Soil Sci 52(4): 314–331

Wischmeier WH, Smith DD (1978) Predicting rainfall erosion losses: a guide to conservation planning. In: Department of agriculture, vol No. 537. Science and Education Administration

Pricing of Carbon Sequestration and Environmental Regulation

4

4.1 Pricing of Carbon Sequestration and GHGs Emission Mitigation

Carbon sequestration in agriculture refers to the storing of carbon for relatively longer time period (10–10,000 years) in plant and soil systems. Plant system includes perennial crops (in agroforestry, silviculture and horticulture systems) which stores carbon in their woody biomass and their well-distributed root systems while soil system includes both inorganic (calcium carbonate, calcareous soil, carbon-rich inorganic parent materials, etc.) and organic carbon (soil organic carbon pools, long-term microbial deposits, wetlands, long-term litter deposition, etc.). Basically, carbon flows/exchanges must be estimated in agriculture for valuation. Carbon flows/exchanges are the balance sheets of carbon inflows and carbon emission. Carbon emission specifically quantified by the carbon equivalent of GHGs (methane, nitrous oxide, carbon dioxide) emissions.

$$\text{Carbon flows} = \text{carbon inflows} - \text{carbon emissions}$$

In agricultural systems, carbon inputs into the soil include stubbles (crop residue remaining in the field after harvesting), roots and rhizodeposition. In wetland, algal biomass addition and in intensive crop management system, addition of manure (FYM, compost, vermicompost, poultry manure, etc.) contributes significantly to carbon sequestration. For example, in rice, 2.5, 19 and 15% of total biomass proportion can be taken for stubble, roots and rhizodeposition, respectively. Then knowing the carbon content of the carbonaceous inputs added to the system, we can quantify the actual carbon sequestered into the soil. As for example, on an average 40% is taken as carbon content of crop residues of annual crop. And carbon leftover in soil could be taken as 28.8% of total added into the system (Mandal et al. 2008).

Life cycle GHGs emissions tools can be used to estimate carbon equivalent emission in agriculture. In this approach, either direct measurements of GHGs

P. Bhattacharyya et al., *Pricing of Ecosystem Services in Agriculture: A Basis of Crop Insurance*, https://doi.org/10.1007/978-981-19-4416-1_4

(CH_4, N_2O and CO_2) emissions could be done or emissions could be estimated by using 'emission factors' for a particular region. In 'emission factor' methods input uses like amount of N-fertilizer and/or manure applied in the field are also taken into consideration (Yan et al. 2003, 2015; IPCC 2006; Hillier et al. 2009; Dubey and Lal 2009). Simulation models (DNDC, INFOCROP, DSAT, et.) also can be used to estimate the amount of GHGs emission in agriculture by using secondary data of climate, soil, irrigation, fertilization, manure, ponding-water-depth and other inter-cultural operation. Carbon emission during agrochemical manufacturing processes, transportation, and farm-gate to consumption (preserving, transporting and storing) must be included in life cycles GHGs emission. After getting the total GHGs emissions of a particular agricultural system through LCA, those could be converted into carbon equivalent emission by using GWP (global warming potential) of respective GHGs (like CH_4, N_2O and CO_2). The GWP of CH_4 is taken as 27.2 (non-fossil origin), whereas the GWP of N_2O is considered as 273 (100 years time horizon) (IPCC 2021). Carbon equivalent emission (CEE) can be estimated by multiplying GWP with equivalent amount of carbon present in CO_2(44/12). Finally, pricing of the carbon flow can be done by multiplying the net carbon flows/exchange with carbon credit (US\$ 21.7 t^{-1} carbon; Bhola and Malhotra 2014; IPCC 2019).

$$\text{Pricing of carbon flows} = (\text{carbon input} - \text{carbon emission}) \times \text{carbon credit value}$$

where

$$\text{Carbon inputs} = \text{stubble} + \text{root} + \text{rhizodeposition} + \text{manure} + \text{algal deposition}$$

$$\text{Carbon } \textit{equivalent emission} = \text{GWP of total GHGs emission} * 44/12$$

$$\text{CEE} = [\text{CO}_2 + (\text{CH}_4 * 27.2) + (\text{N}_2\text{O} * 273)]\ \text{CO}_2 \text{ equivalent emission} * 44/12]$$

This pricing is specific to a particular agricultural system. And the GWP values can be modified according to the life span and the objective of the study.

The carbon sequestration of a system (including plant and soil) often expressed as net carbon flow/exchanges. However, in that case time period of carbon storage should be considered for larger period. At least 10 years for annual agriculture crops, perennial horticultural, silviculture and agroforestry plantation.

The 'bottom-up' approach is an effective tool to quantify the economic value of ES in agriculture. Here, the pricing of ES is done at the field level based on experimental data (in contrast of value transfer system) and with sufficient replication then extrapolate to a larger scale considering the land use pattern. In this approach, carbon accumulation is considered as 'offset' of the carbon dioxide emission/GHGs emission to the atmosphere.

Soil organic carbon (SOC) storage (accumulation) in agriculture can alternatively be estimated by balancing carbon input (in form of soil organic matter) and output.

The SOC inputs include litter deposition, manure application, roots, stubbles (residues leftover in field after harvest) and rhizodeposition (including root exudates), whereas two major SOC outputs are CO_2 and CH_4 emissions and carbon removal through harvest. Therefore, SOC storage can be estimated by multiplying unit price replacement of SOC to the SOC accumulation/storage in a particular agricultural system.

$$A_{SOC} = A_{mc} + A_{rl} + A_r + A_{rd} - A_{CH4} - A_{CO2}$$

$$Vsoc = A_{soc}\ X\ P_{soc/som}$$

where

A_{soc} = Amount of SOC storage/accumulated in agriculture soil (kg ha^{-1}y^{-1} as carbon).

A_{mc} = Amount of carbon added through manure (kg ha^{-1}y^{-1} as carbon; 11–20% of total aboveground biomass).

A_{rl} = Amount of organic material leftover added (kg ha^{-1}y^{-1} as carbon measured or 20–30% of above-ground biomass).

A_r = Amount of root biomass added (kg ha^{-1}y^{-1} as carbon measured).

A_{rd} = Amount of rhizodeposition (kg ha^{-1}y^{-1} as carbon; generally, 3–4 times of root biomass).

A_{CH4} = Methane emission (kg ha^{-1}y^{-1}, as CO_2 equivalent and compared to carbon).

A_{CO2} = Carbon dioxide emission (kg ha^{-1}y^{-1} as carbon).

V_{SOC} = Price of SOC storage/accumulation in agriculture (US$ per kg per hectare per year as carbon).

$P_{SOC/SOM}$ = Replacement price of SOC (US$ per kg of carbon).

(Watanabe and Roger 1985; Yu et al. 2011)

4.1.1 Carbon Sink in Wetlands

4.1.1.1 Rice Paddy Systems

The importance of natural wetland and anthropogenic wetlands (rice paddies) for providing/maintaining ESs in agriculture is increasing day by day with the decreasing trend of forest and grassland for industries and urban use. The economic valuation of ESs provided by rice paddies in terms of gas regulation, soil organic carbon sequestration, water regulation, nitrogen fixation and flood control apart from primary production become important. Anthropogenic wetland (rice paddies), the largest human control ecosystem (~1.5×10^8 ha globally; producing 28–30% world grain production) provides significant ESs like pollination, habitat to birds, beneficial insects, flood control, water and gas regulation, nutrient and carbon cycling, aesthetic, etc. (Wang and Jiang 2021; Muto and Yokokawa 2022). However, there are some disservices also through greenhouse gases (GHGs) emission and excessive use of synthetic chemical fertilizers and pesticides. On an average, agriculture

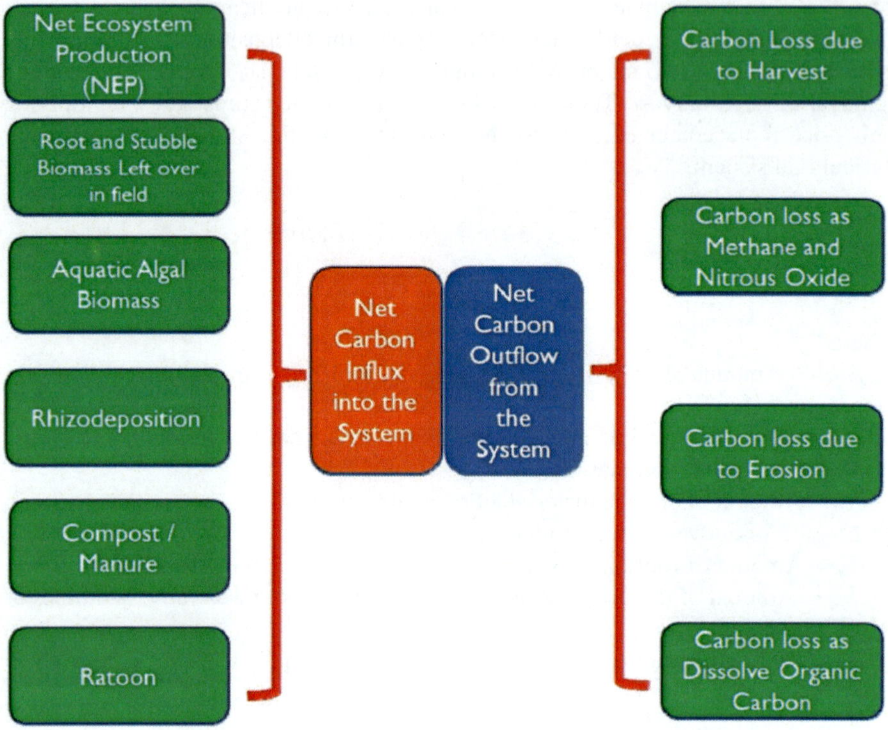

Fig. 4.1 Carbon balance of rice paddy system based on eddy covariance techniques

(including rice paddies) and natural wetlands contribute 29–34% of total methane (CH_4) and 60% of total nitrous dioxide (N_2O) emission globally (IPCC 2007, 2019). On the other hand, rice plants assimilate huge amount of CO_2 and emits O_2 through photosynthesis. Rice paddies provide huge ESs which include assimilation of carbon in above-ground plant parts, roots, adding carbon to soil through rhizodeposition; recharging groundwater (as most of the rice paddies are cultivated in bunded condition for maintaining standing water in field for 100–120 days per year); providing nest to many beneficial insects, predators of pest, soil-microorganism; fixing nitrogen biologically through bacteria and algae in the system, etc. An account of carbon balance in lowland rice ecology in Eastern-India showed that rice paddy acts as net carbon sink when considered all possible carbon input and output including GHGs emissions (by conversion of CH_4 and N_2O into CO_2 equivalent) (Fig. 4.1) (Bhattacharyya et al. 2014; Swain et al. 2016). The real-time measurement employing advanced eddy covariance technique was used for estimating the carbon balance. The following carbon balance equation can be used for carbon credit compliance:

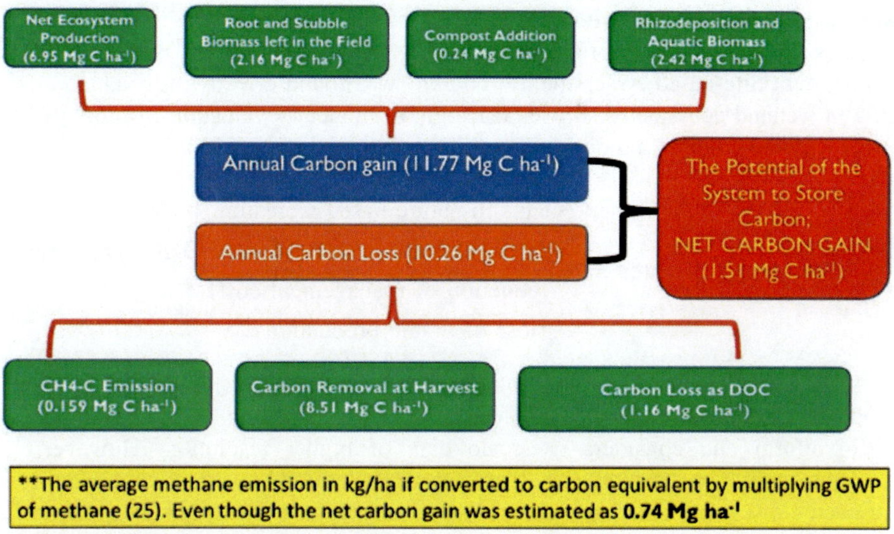

Fig. 4.2 Quantified data of net carbon balance of rice paddy system based on eddy covariance techniques (Adapted from Bhattacharyya et al. 2014)

$$\text{NECB} = \text{NEP} - \text{D} - \text{F} - \text{H} - \text{VOC} - \text{NEME} - \text{N} - \text{E} + \text{I}$$

where, NEP: Net Ecosystem Production; D: Dissolved Organic Carbon; F: Loss of carbon by fire; H: Loss of carbon during harvesting; VOC: Loss of carbon as Volatile Organic Compound; NEME: Loss of carbon as methane after converting CO_2 equivalent by GWP; N: Loss of carbon due to N_2O emissions (as CO_2 equivalent by GWP, N_2O); E: Loos of carbon by erosion and eluviation; I: Addition of carbon from litters, organic manure and other organic compounds.

Overall, the lowland rice paddy system could sink carbon @ 1.04 Mg ha^{-1} $yr.^{-1}$. Now if we convert it to monetary value keeping US$ 40 per ton of carbon sequestration (as carbon credit) then it would be around US$ 41.6 ha^{-1} annually. Now taking 10–14 million hectares (average 12 m ha) land in India grows rice in lowland (rice paddy) condition, so we can estimate the system has the potential to provide ~U $ 500 million annually as ESs, through carbon storing only (Bhattacharyya et al. 2014; Swain et al. 2016). This priced ESs should be counted as regulatory/ supporting ESs through rice/agriculture. Rai et al. (2018) also accounted an increase of 0.17–0.20 Mg ha^{-1} y^{-1} carbon (0.34–0.41 Mg ha^{-1} y^{-1} soil organic matter). Here, also in terms of carbon credit, the US$ 8 $ha^{-1}y^{-1}$ (based on 2010 carbon credit value in $) could be priced for storing the soil carbon only.

Another quantification of net carbon balance in lowland rice ecologies has been reported by Bhattacharyya et al. (2014) of two season rice separately using eddy covariance techniques considering GWP of methane and converted the emission in carbon equivalent (Fig. 4.2).

4.1.1.2 Models and Toolkit for Pricing of ESs in Wetlands

Models and toolkits are useful for pricing ESs in wetlands. In this regard, TESSA (Ecosystem Site-based Assessment) 'Toolkit' was found effective for valuation of ESs in wetland ecosystems. It uses the logic of monetary valuation of biophysical processes and compared the former arable land to 'restored wetland'.

$$\textbf{Restoration of wetland} = \begin{array}{l}(\text{Gains in nature based recreation} + \text{Grazing} \\ + \text{flood protection} + \text{reduction of GHGs emission} \\ + \text{reduction of management cost}) \\ - (\text{loss of arable production})\end{array}$$

In the simple term, TEESA tool kit compares the prices of ESs of restored wetland to that of adjacent arable land. It uses specific technical logic in GIS-platform and considers site-scale data of higher resolution. However, it accounts only five specific ESs like water regulation, harvested wild-products, cultivated-products, nature-based regulation, and climate regulation. For example, around US\$200 ha^{-1} year^{-1} could be the net value of ESs resulting for conversion of arable crop land to 'restored wetland' (taken US\$22.75 t^{-1} CO_2) (Richmond et al. 2007; Bryan and Crossman 2013). The tool kit allows to assess the GHGs fluxes (CO_2, CH_4 and N_2O) under different land uses and can make use of both primary (measured data) and secondary published data on emissions. Flux conversion to net CO_2-equivalent emissions is done by using GWP of CH_4 and N_2O, which are the part of the 'toolkit' calculation framework. Ultimately, pricing GHGs fluxes are calculated considering the local market price of carbon.

4.1.1.3 Peatland and Sea Grass

Peatland plays a key role in terrestrial carbon sequestration. Protection and promotion of peatland was a major issue in 'Kyoto Protocol', 'Ramsar convention' and 'UN convention on Biological Diversity'. Peatlands significantly contribute to climate change regulation in terms of carbon sequestration, biodiversity maintenance, water recharging and recreational benefits (Climate Change 2014; IPCC 2014; Reed et al. 2016). Anoxic soil condition, poor drainage, continuous water logging and slow organic carbon decomposition rate make peatland an ideal place for storing carbon for long time. Healthy peatland has been storing carbon for 10,000 years and has a considerable cooling impact on global climate. However, when drained this organic carbon rich ecology contributes significantly towards GHGs sources. Estimates revealed that up to 25% global CO_2 emissions from land use sectors contributed by degraded peatlands having a high social and environmental cost. Converting back the present crop land, pasture and degraded-peatland again to healthy peatland by rewetting, restoration through 'Sphagnum-growth', and flood-regulation are effective mitigation options to climate change. Encouragingly, there are growing interest for development of regional/local market for peatland restoration in various countries including the UK, the USA and Germany. Restoration of peatland found to be a relatively cost-effective mitigation options having good abatement potential. Now, pricing of ESs provided by peatland can be indirectly done through restoration cost of degraded land with specific time frame. In this aspect, regional carbon market having transparent accreditation and verification protocols with precise GHGs emission estimation methodologies considering both investor, service provider and managers are necessary. Transparent cost-effective, robust assessment and monitoring of GHGs emissions is a challenge to develop 'regional carbon market'. So, economic incentives for restoration activities through 'agri-environment' schemes should be encouraged. Scientific evidence-based ecosystem service markets for wetland/peatland restoration are needed. Global standard-based carbon markets should be extended at regional/local level with proper linking to ESs markets.

Like peatland, seagrass ecosystems in wetland also provide several ESs. Few ESs provided by seagrass systems include coastal bank protection, coastal erosion control, maintaining diversities of marine life and fisheries, and 'blue carbon' sequestration. Seagrass system provides all four components of ESs, namely regulating, supporting, provisioning and cultural services. This ecosystem is responsible for 3–20% of carbon sequestration in marine systems, globally (76–151 Tg carbon; @6–7 t carbon ha^{-1} $year^{-1}$) (Duarte et al. 2013; IPCC 2019). Seagrasses facilitate to develop deep (even up to 11 m depth) organic carbon rich sediment in wetlands of marine system. However, estimate indicated that 60–300 Tg carbon $year^{-1}$ can be lost as CO_2 to atmosphere due to loss of seagrass biomass (Fourqurean et al. 2012). At the same time, biomass loss of seagrass also reduces the carbon accumulation rate in sediments. Present trends of seagrass loss could be accounted for 130–520 Mg CO_2 losses per hectare which is around 10% of carbon that emitted through land use change (LUC), annually in the world. But the prospect of developing 'carbon credit' markets or 'ES-market' for seagrass restoration is still slim. However, voluntary carbon market has some opportunities but faced challenges from carbon-price fluctuations, carbon-benefits quantification and knowledge gaps on 'blue carbon science'.

4.1.1.4 Rainwater Harvesting Ponds

Rainwater harvesting ponds (RHP) and natural village ponds are the anthropogenic wetlands provide ESs in term of carbon sequestration, regulation of hydrology, maintenance of biodiversities and cultural services. Emergent vegetation and organic carbon deposition in sediments are crucial for carbon sequestration in ponds. However, methane emissions or methane trapping in the ponds contributed to net carbon sequestration in the system. Now, we will discuss how the carbon sequestration can be estimated in ponds (RHP and others) and subsequently how the ES could be valued. Firstly, organic carbon content of ponds/wetlands is measured and correlated with the age of the pond. Generally, 10–15 cm upper sediment is collected through soil-core. The sampling must be done randomly (with replications) in higher frequency representing each hydrological unit. Depth of sediment sample can be varied from site to site based on pond type, run-off characteristics, erosion status, etc. A separate soil-core should also be used to collect the samples for estimation of bulk density of sediment/soil in each sampling site. The total carbon contents of the sediment are measured by dry combustion method. Now the carbon density is estimated by the following equation.

$$\begin{aligned} &\text{Carbon density } \left(\text{g carbon } m^{-2}\right) \text{of each pond} \\ &\quad = \%\text{carbon content of soil or sediment} \\ &\qquad \times \text{bulk density of the respective sample} \times \text{depth of soil or sediment sample} \end{aligned}$$

Then, to quantify the sequestration rate, the carbon densities of each pond/zone are regressed with pond age. Now, multiplying the carbon credit value with the amount of carbon sequestrated per unit area annually in a specific location/region the pricing of carbon sequestration can be calculated. Values of other ESs of ponds like biodiversity maintenance and cultural services should be added with carbon sequestration values to get total values/prices of ESs provided by ponds.

4.1.2 Carbon Sequestration in Agroforestry, Silviculture and Horticulture

In agroforestry, only a part of arable land is used for trees. So, provisional services of arable crops rarely hampered, rather yield of crops sometime increased in agroforestry system. Agroforestry is a profitable proposition with good carbon sequestration potential if managed properly. However, in 'Clean Development Mechanism (CDM)' has not considered incentives for afforestation and restoration of agroforestry system, separately, which hinders the large-scale implementation of agroforestry system. So, to implement agroforestry in large scale for climate change mitigation need incentives at policy level. For that proper understanding of supply cost of sequestration in the context of socio-economic perspective is needed. Specifically, carbon sequestration cost of agroforestry system can be calculated by 'partial market equilibrium' considering the 'average cost curve' and 'economic break analyses' to fix the 'supply cost'. There are two primary cost components, (i) implementation cost and (ii) opportunity cost. The average cost curve can be drawn by average cost of sequestration ($ per ton of carbon per year) as a function of area (hectare) at which agroforestry has been done or the sequestration potential of the area (ton carbon). This type of approach encourages private landowners and NGOs to participate in the implementation of agroforestry projects. To nullify the regional fluctuation of carbon price, the average cost curve can be generated as function of potential carbon price due to sequestration. Lesson learned for site-specific studies, that lower 'transaction cost', higher 'carbon payments' and 'invest at high quality baseline', facilitates the promotion of agroforestry, silviculture and perennial horticultural systems.

Based on the 'Break even analysis', the minimum extent area for a profitable agroforestry project can be calculated by the following equation:

$$A = \frac{C_f}{(P_c \times C_s) - C_v}$$

where

A = area of agroforestry implemented for carbon sequestration (ha).

C_f = fixed cost ($).

Pc = carbon price ($ per ton of carbon).

Cs = carbon sequestered (ton carbon per hectare).

Cv = viable cost ($/ha).

It must be remembered that the marginal income per hectare ($Pc \times Cs$) should be higher than the viable cost (Cv) for a viable and sustainable agroforestry system. However, in an agroforestry or silviculture or perennial horticultural system 'Average net present value (ANPV)' per hectare per year is the key to judge the viability and sustainability of the system. The ANPV can be calculated by the following equation:

$$ANPV = \frac{1}{T} \times \sum_{t=0}^{T} d[PC_t \times CS_t] - \frac{1}{T} \times \sum_{t=0}^{T} d\left[\frac{(IC + TC)}{A} + Iv + Tv + Oct\right]$$

where

$ANPV$ = average net present value ($ ha^{-1} year^{-1}).

t = time in year.

T = total duration of the project/scheme (years).

d = discount factor.

Pc_t = carbon price at time 't' ($/ton carbon).

CS_t = carbon sequestration rate 't-1' years to 't' years (ton carbon/ha).

IC = implementation fixed cost ($) in year 't'.

TC = transaction fixed cost in year 't' ($).

Iv = implementation variable cost ($/ha) in year 't'.

Tv = Transaction variable cost in year 't' ($/ha).

Oct = opportunity cost in year 't' ($/ha).

A = area of agroforestry system (ha).

The carbon sequestration can be estimated for 10 to 100 years in agroforestry system with spatial variation. However, it varies with plant type and region of the plantation.

Similar approaches could be used for pricing of silviculture and perennial horticultural system for pricing of carbon sequestration as a part of valuation of ESs of the system. The general observation is that carbon supply is most cost-efficient in highly productive silviculture system. A reduction of carbon price can be observed due to increase in wood price in silvicultural system. So, there is a 'trade-off' between provisional and regulatory ESs in this system which need to be tackled through policy framework. The deterministic approach is more useful than static approach in agroforestry/silviculture systems where woods are used for commercial benefits. Here, marginal carbon sequestration cost derived from '*Faustmann's Formula*' could be used for calculating carbon supply from the opportunity cost of the system (agroforestry/silviculture/horticulture, etc.). Here, marginal abatement cost represents the lowest cost for reducing GHGs emission by an additional unit also could be considered to promote the systems having higher carbon sequestration potential.

4.1.3 Gas Regulation and GHGs Emissions

Gas regulation in respect to agriculture can be classified into two broad categories. One is to emit oxygen and absorb CO_2 (as sink) and another for emitting GHGs (CO_2, CH_4 and NO_2) as source during respiration and other metabolic biochemical processes.

Oxygen emission and its economic value can be calculated by:

$$Q_0 = 1.1g.m_{npp} - q_{so}$$

$$V_0 = Q_0 - P_0$$

where

Q_0 = amount/quality of O_2 regulation (Mg ha^{-1} y^{-1}).

1.1 g = Coefficient of Net Primary Production (Npp) to O_2 (for Rice)m_{npp} = net primary production (Mg ha^{-1} y^{-1}).

q_{so} = O_2 consumption by plant system (Mg ha^{-1} y^{-1}).

V_0 = price (economic value) of O_2 regulation (USD ha^{-1} y^{-1}).

P_0 = replacement price of O_2 (USD kg^{-1}).

(Wang et al. 2010)

Further, the GHGs regulation in agriculture and its pricing can be calculated by the following empirical equations:

$$Q_g = (1.63.m_{npp} - q_{so}) - (GWP\ of\ CH_4.q_m + GWP\ of\ N_2O.qne)$$

$$V_g = Q_p - P_g$$

where

Q_g = amount/quantity of GHGs regulation in agriculture (ref. to rice paddy) (Mg ha^{-1} y^{-1}).

1.63 = Coefficient of Net Primary Production to CO_2 Assimilates; in Reference to Rice but Can Vary Crop to Crop; q_{so} = CO_2 emission by plants/organic in soil (Mg ha^{-1} y^{-1}).

GWP of CH_4 = 27.2 for 100 years (IPCC, AR6 2021).

GWP of N_2O = 273 for 100 years (IPCC, AR6 2021).

q_m = CH_4 emission (Mg ha^{-1} y^{-1}).

q_{ne} = N_2O emission (Mg ha^{-1} y^{-1}).

V_g = pricing/economic value of GHGs regulation (USD ha^{-1} y^{-1}).

P_g = cost of carbon credit (US$ 40t^{-1} of carbon); this could be varied county to county or region to region or can be taken from consented global platform.

The most economically efficient approach to mitigate GHGs emissions is to reduce the GHGs emission to a particular point so that the marginal benefits of the reduction should equal to its marginal cost. That means damages caused by emitting an addition unit of GHGs (or CO_2-equivalent) should be monetized to fix the carbon tax rate for getting equivalent marginal benefit of the emission reduction. So, by this way externality of carbon saving benefit can be internalized in the market system which is a cost-effective approach to mitigate GHGs emission. However, most of the climate policies targeted on the 'supply-side' restriction approach like mandatory biofuel blending with fossil fuel, subsidizing renewable energy source production (solar, wind power, etc.), etc. But there is a need to fixing price for adopting a specific emission mitigation technology or generating the new product. Major challenges to fix the GHGs emissions reduction costs are high initial cost for infrastructure development; political; lack of consideration of deterministic dynamic cost involved in the processes. Therefore, from economic point of view the right incentives for emission reduction rather than subsidies to the technologies which could produce low carbon energy sources is important for quick adaptation of GHGs mitigation technologies in agriculture.

4.2 Pricing of Bioremediation for Pest and Pollution Control

Bioremediation is the green technology that involves organic substance, microbes and tolerant crops (phytoremediation) to clean up the groundwater, air and/or soil from toxic pollutant and heavy metals. Specifically, certain plants have unique mechanisms in their absorption, exclusion and transport systems which selectively remove the heavy metals and toxic contaminants from the soil and water. However, initial infrastructural cost of bioremediation and delayed response on remediation benefits compared to physical and chemical treatments of pollutants in soil and water impose hurdle of their (bioremediation practices) adaptation on large scale. Therefore, long-term impact of bioremediation in terms of removing/sequestering environmental pollutants and providing ecosystem services must be valued for 'adopting this green-technologies'.

Heavy metal pollution is a serious issue to the environmental health. It is a global problem but appears primarily sporadically in specific region of the world. Heavy metal pollution not only reduces crop yield but also causes health problem of soil, animal and human being and provides ecosystem disservices. The clean-up cost of heavy metal contamination in soil through chemical and mechanical means is high. In this aspect, bioremediation plays an important role which is economic and environment friendly. Pricing of bioremediation should be done in terms of pollution control, reducing hazards and providing ecosystem services. So, it is necessary for giving these practices an economic outlook, which is helpful for future planning of policymakers. Pricing of ecosystem services (ESs) provided by bioremediation of heavy metal pollution in soil would encourage the better allocation of remediation funds in this arena. Therefore, suitable methods and quantified assessment are needed. A well-accepted method by Wang et al. (2016) is briefly discussed in the next paragraph.

Bioremediation cost can be divided into capital and operational cost. Capital cost includes initial pollution assessment, equipment procurement, sample processing and infrastructure facility (road, store house, building, etc.) development. On the other hand, operational cost includes cost of planting materials, labour, custom hiring of large machines, maintenance of planting materials, cost of intercultural operations (seeds, fertilizers, land preparation, fencing, weeding, pest control, harvesting) and indirect cost (power, electricity, fuel, etc.). Now benefit of bioremediation and soil pollution control practices should be considered after recovery (remediation), based on how the soil functions, health improves, and reduction of health-related threats and other ecosystem services. So, pricing needs to be calculated indirectly by the quantity of yield decrease, reduced ESs functions, ill-effect of human, animals and soil health if remediation has not been done. As the beneficial effects would be long lasting, so future benefit also should be considered.

So, gain of bioremediation can be empirically written as:

$$BR_T = BR_P + BR_F$$

where BR_T refers to total gain due to bioremediation; $BR_P =$ present gains during remediation; $BR_F =$ future gains after remediation.

$$BR_P = BR_{PV} \times BR_{PY}$$

where $BR_{PV} =$ present value of crops; $BR_{PY} =$ present yield of crops.

$$BR_F = BR_{FY} + BR_{FS} + BR_{FH}$$

where $BR_{FY} =$ Gains due to recovering the functions of soil to produce desirable/potential yield of that system; $BR_{FS} =$ Gains due to receiving soil as a healthy system to function properly; $\mathrm{BR_{FH}} =$ Gains due to preventing human income loss.

Further, BR_{FS} and BR_{FH} can be taken as just-for-one time benefit, but $\mathrm{BR_{FY}}$ should be based on annual/bi-annual benefit according to the duration of crops would be cultivated in the remediated soil/land.

$$BR_{FY} = \sum P_i \times Y_i \times \Delta A_i$$

where $Pi =$ price of crops/products. $Yi =$ yield of the crops/products. $\Delta\ Ai =$ the increase in soil/land area due to bioremediation.

The $\mathrm{BR_{FS}}$ includes the gains/benefits of preventing pollution from soil to air and water, erosion control, and checking biodiversity loss. Whereas, the $\mathrm{BR_{FH}}$ could be calculated by the sum of gains due to reduced hospitalization cost and the reduction in individual income caused by early-death/disease.

Based on the above-mentioned approach, Wang et al. (2016) reported that ecosystem service functions could benefit US$1015 per square hectometre annually. The study was based on two-year bioremediation data where soils were contaminated with cadmium, arsenic and lead. They also concluded that the gains of bioremediation are expected to offset the implementation cost by less than 7 years (Wang et al. 2016).

In this aspect, efficiency of bio-/phytoremediation through naturally available plants/organism is a limitation. Often the efficacy of natural bioremediation is less due to low biomass production and limited sequestration potential. To address this issue, the genetically modified plants ('green-plant technology') could be effective which is low-cost, non-intrusive and quick-responsive. But the public opposition to genetically modified plant is a concern. However, the ecosystem benefits of bioremediation to restore the balance of a 'stress environment,' seem to far outweigh its cost (Paz-Alberto and Sigua 2013; Melinda 2013).

4.2.1 Pricing of Biological Control of Pests

These services are quantified indirectly by estimating the 'predation rate', specifically by quantifying the predators and parasitoids for a particular field in an agricultural system (Sandhu and Heinrich 2005; Van Mele 2008). Here 'economic threshold level' (ETL) (ETL refers to the limit at which farmers are recommended to apply pesticides) is generally used to judge the pest infestation rate per hectare (unit area). Based on that the recommended dose of pesticide per hectare (required to control the pest) is multiplied with the market price of that pesticide and added with the application cost to get the price (cost) of pesticide control per hectare (unit area). Subsequently, 'predation rate' is used to estimate the total number of pests controlled or removed. Then the value of biological control of pests is calculated by multiplying the total number of pests controlled by the cost involved in controlling the respective pests (Nayak et al. 2019).

$$\text{Pricing of biological control of pest} = \text{Total no.}$$
$$\text{of pest controlled} \times \text{Cost of pesticides for controlling the respective pests}$$

4.3 Pricing of Waste water Utilization in Agriculture

Benefits of waste water utilization in agriculture are many facets. Benefits include energy saving for pumping fresh water; saving of freshwater reserve; fertilizer saving; less of native soil phosphorous extraction by plants; and reduction of carbon emission to atmosphere. Additionally, use of waste water could enhance crop yield and promote more economic cropping sequences. However, waste water should be used after proper treatment as it contains toxic microbes and chemicals that could pose risk on human and animal health. Waste water can be used effectively in water-cooling, flushing of toilet, groundwater recharging, making urban parks, recreational ponds, industrial wash that helps in reducing carbon footprint on those sectors and indirectly providing ESs. In agriculture, waste water irrigation (obeying irrigation safety norms) is beneficial for improving crop yield and reducing water shortage. It also helps in restoring inherent soil fertility and reducing fertilizer use cost by supplying nutrients to crops. Waste water plays an important role in reducing inland soil salinity and by virtue provides supporting ESs. The use of waste water irrigation in agroforestry and degraded land plantation found safe and beneficial. However, no specific valuation technique is available for pricing ESs provided by waste water utilization in agriculture. But adopting the waste water utilization plan in mainstream of future water governance would definitely encourage economist and scientist to develop precise methodologies for valuation of waste water utilization in agriculture.

4.4 Valuation of Residue and Its Management Options

4.4.1 Challenges and Potential of Crop Residue Management

The major challenges are that the growth in crop residue (particularly rice straw) generation and the supporting infrastructure for its management are not moving at the same pace which cause straw burning that lead to environmental pollution, GHGs emissions and climate change. The structural, institutional, technical and socio-economic challenges are associated with residue management that results into the residue burning globally, and India in specific. In parts of India, open field straw burning is a compulsion for farmers rather than a choice. However, the turning of the wheel is possible. Around 16% of crop residues are subjected to burn on field in India out of which 60% is rice straw (Bhattacharyya et al. 2020). However, we can make this waste (surplus rice straw) to wealth by utilizing it in bioethanol production, biochar conversion, mushroom production, composting, pulp-making, fuel-briquette preparation and conservation agriculture (CA) based on the site-specific demands and potentialities.

Rice straw burning has significant negative consequences on environmental quality. It pollutes air and causes the emissions of greenhouse gases (GHGs). Open field residue burning emits toxic gases and particulate matters. Toxic gases include sulphur dioxide (SO_2), carbon monoxide (CO), nitrogen oxides (NOx), volatile organic compounds (VOC), gaseous hydrochloric acid (HCl), dioxins, furans, carcinogenic polycyclic aromatic hydrocarbons (PAH), etc. Further, residue burning considerably contributes to suspended particles (fine particles ($PM_{2.5}$) and coarse dust particles (PM_{10})) which are deteriorating the air quality. Rice straw burning causes GHGs emissions. Estimates revealed that around 700–4100 mg of methane and 19–57 mg of nitrous oxide can be generated on open field burning of one kilogram of dry straw (Kritee et al. 2018; Gerber et al. 2016). Recent reports depicted that the northern Indian states have been seriously affected by air pollution in last 2–3 decades due to straw burning (Bhattacharyya et al. 2021). Black carbon and methane are the short-lived climate pollutants released directly from straw

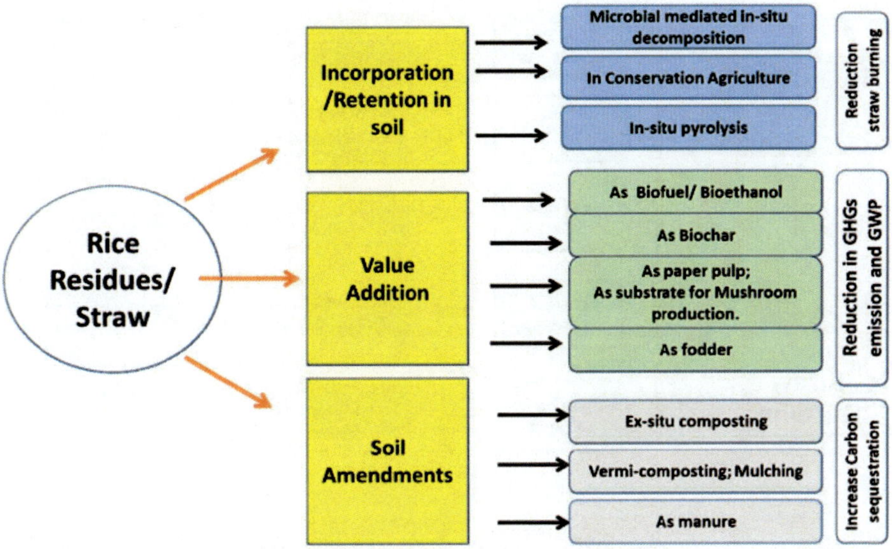

Fig. 4.3 The different options of utilization of rice straw in agriculture

burning. Apart from those a significant quantity of nutrients is lost because of straw burning. For example, nitrogen, potassium and sulphur are lost around 40, 33 and 45%, respectively due to open field residue burning (Bhattacharyya et al. 2020).

However, there are good potential to utilize crop residue in general and rice straw in particular for several profitable avenues. For example, rice straw, the bulkiest crop residue generated in world (731 m t) could be effectively utilized as animal feed; bioethanol, biogas, biofuel production; cellulosic feed-material for paper-pulp industries; mushroom production; manures and composting; soil conservation; etc. (Fig. 4.3). Conservation agriculture and mulching are other two eco-friendly management practices of crop residues in agriculture. However, for adaptation of conservation agricultural practices on large-scale proper synergy of second-generation straw handling machines (e.g. Happy Seeder, Paddy straw Chopper cum Spreader, Straw Management System, etc.) with real-time maintenance and advisory services are necessary.

4.4.2 Pricing of Alternative Options of Residue Management

Environmental and economic gains and losses of alternate straw utilization practices over field burning includes tangible and intangible cost. For example, in case of rice straw utilization, straw-retention could be taken as baseline and it can be compared with other straw-management options like biochar generation, open field burning, bioethanol conversion, small briquette production for fuel, animal feed and conservation agricultural practices. One basic land unit with certain residue producing

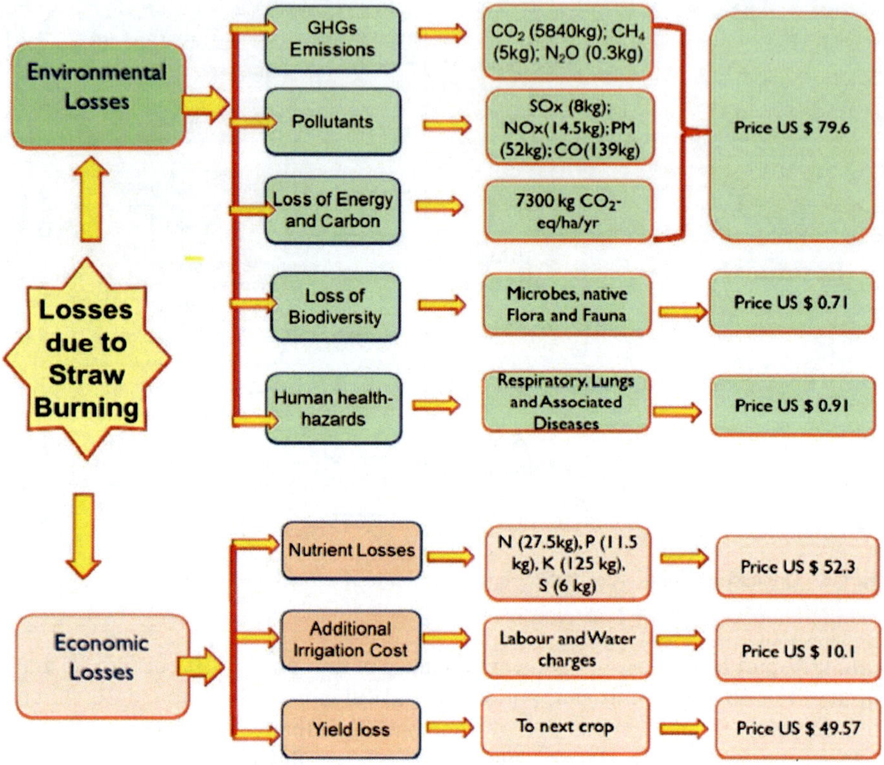

Fig. 4.4 Pricing of environmental and economic losses of rice straw burning. [Note: All estimation is made per hectare per year basis; value of US$ 2021 was taken for calculation; Based on both measured and published data; Sources: GOI 2013; Kumar et al. 2015; NAAS 2017; Bhattacharyya et al. 2021]

capacity should be taken on annual basis (e.g. one-hectare land area having 5 tonnes straw producing capacity per annuum).

The first steps of pricing of losses/gains are to categorize the processes/items and list them. Then quantify the gains/losses of each process/item per unit area per annum. After that the quantified gains/losses are multiplied by unit price of that commodity. The unit price could be taken from market value if market is available or can be indirectly estimated from 'offsets' which is correlated with those and can be valued. For example, the economic losses due to rice straw burning could be categorized as: (i) nutrients losses; (ii) yield losses to the next crop; (iii) irrigation charges; (iv) emissions of GHGs and toxic pollutants; (v) biodiversity losses and (vi) human health hazards (Kumar et al. 2015; Bhattacharyya et al. 2021) (Fig. 4.4). Some of them are tangible, like nutrients losses, which can be calculated through the opportunity costs by leaving the crop residues as fertilizer and then estimate it by fertilizer replacement cost (NAAS 2017). Pricing of irrigation could be done with the soil moisture status, labour and water-conveyance charges. Straw burning is associated with intangible losses like soil health, biodiversity and human health. Those are environmental losses and an opportunity cost linked with those. For example, the cost of treatment of human health hazards due to burning can be taken as ~ INR 4.52 per capita welfare loss (Kumar et al. 2015) that can be converted into per unit area (hectare) losses of welfare. A detailed example of losses due to straw burning are presented with some economic values for easy understanding in Fig. 4.4 (Suresh et al. 2021; Bhattacharyya et al. 2021). It is important to understand that there are no environmental gains as per rice straw open field burning is concerned. However, some researchers argued that to consider the transportation charges (as gains) of straw (from field to factory) for other alternative commercial uses in industries. So, if we put all the tangible and intangible gains/losses of a particular processes/management practices of reissues per unit area annually, we can have the balance sheet and can easily calculate the pricing of straw management practices on economic terms. On those balance sheets, the environmental gains can easily be considered as ecosystem services provided by certain residue management practices.

The environmental and economic assessment model given by Ji et al. (2018) can be used for pricing the intangible gains obtained from biochar and fuel-briquette generation from rice straw. The same approach could be used for assessing the ecosystem services provided by those processes compared to straw burning and straw-retention in the field. The model estimated the GHGs emissions for biochar generation in the tune of 0.75-ton CO_2-eq per hectare causes the environmental losses. These losses were estimated by considering the emissions during infrastructure development, equipment-manufacturing, straw-collection and transportation; and carbon emission due to nutrient losses during biochar conversion. The environmental gains which subsequently provide the ecosystem services through these processes includes carbon sequestration, energy displacement and reduction of nitrous oxide emissions (8.75 t CO_2-eq per hectare). Apart from those the biochar conversion fetched the economic gains on account of net primary value (NPV) (Ji et al. 2018). The NPV is calculated with the help of cost of capital investment (feedstock), maintenance-charges, labour man-days, taxes and annual revenue through selling of the products. Therefore, the net economic gains for biochar conversion from rice straw were varied from country to country and regions (e.g. US$ 388 in India and US$ 183 in China per hectare) (Bhattacharyya et al. 2021) (Fig. 4.5). Similarly, for fuel-briquette production from straw the environmental gains which would ultimately be translated to ecosystem services primarily come from energy displacement. Now if we want to value the ecosystem services (environmental gains) for conversion of straw to bioethanol, we should consider the fossil fuel-replacement due to use of bioethanol (e.g. net economic gain was US$ 664 in India context based on hydroelectricity; Sreekumar et al. 2020; Soam et al. 2016; Bhattacharyya et al. 2021) (Fig. 4.5). Utilization of crop residues in "conservation agricultural (CA)", fetched environmental gains in terms of GHGs emissions-mitigation, nutrient build-up, conservation of soil and biodiversity (Kumar et al. 2015; Launio et al. 2016). However, it should be noted that getting the services of these processes infrastructure facilities, and support from policy level are necessary.

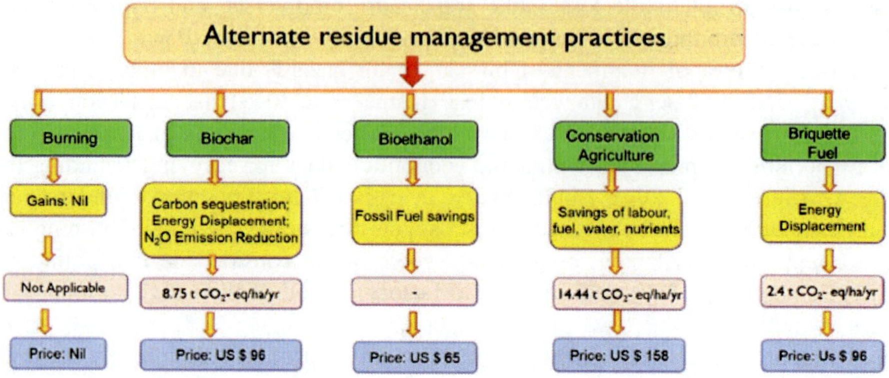

Fig. 4.5 Pricing of environmental gains towards ecosystem services in alternate rice straw management practices. [Note: All estimation is made per hectare (straw production of 5 tonnes) per year basis; value of US$ 2021 was taken for calculation; Based on both measured and published data; Sources: Soam et al. 2016; Ji et al. 2018; Bhattacharyya et al. 2021]

The pricing of ecosystems services in terms of environmental gains for alternate uses of crop residues in agriculture would support the green policies of residue management on long-term basis and building up the necessary infrastructure to support those.

4.5 Future Road Map

4.5.1 Cooperative Approach for Valuation of Ecosystem Services in Agriculture

Majority of the options and methods to value the ESs in agriculture are concentrated around government regulation and/or markets-incentives. However, another approach could be a 'cooperative solution'. In this approach, a collective management based on the resource characterization for providing optimum ES is needed. This approach should have a logical framework of interrelated activities for a particular ES or the flows of ESs at different scale. In this approach, consideration of long-term goal and management-focus should be extended to non-production related ESs (regulatory and supportive services) (Antle and Capablo 2002; Zhang et al. 2007; Adhikari and Hartemink 2016). The 'Cooperative solution' of pricing of ESs in agriculture can be dealt with collective restoration of wetlands, maintenance of natural habitats with social norm and credit sharing and building the cooperation with landowners through long-term benefit agreement (Goldman et al. 2007). Present day 'incentive-structure' in intensive agriculture (either in the form of subsides to inputs or incentive for using certain technologies) builds on the short-term support to the production of food, fibre, fuel and fodder and often at the expense or supportive and regulatory ESs (loss of soil fertility, water table depletion, soil quality deterioration, damage of beneficial insects and pollinators, etc.). Farmers are not really interested to provide these supportive services as they are not getting direct short-term benefits from them. In virtue, many a time disservice to the ecosystem is performed. The 'cooperative solution' could create a mechanism by which farmers can receive the price-premium on services they provided to the society or ecology as a whole. In this approach, collective benefit is to be valued at landscape and regional scale and the contributors to provide those services are incentivized by other who are enjoying the benefits by following a logical framework. However, all ESs are not suited in this approach. Specifically, pollination, pest control, recreation, biodiversity maintenance and aesthetic landscape could be considered in 'cooperative solution'. Services like maintenance of soil fertile, structure, water purification and flood control are generally not suited for valuation in this approach (Stallman 2011). In nutshell, 'command and control' approach either by private or government organization often failed to protect and maintain the 'natural capital' (like ESs in agriculture), but collective management of natural resources provides a 'cooperative solution' which is sustainable for long term as it encourages attachment of the people with the native and mutual benefit and credit sharing through active participation in the processes (Arnald and Campbell 1986; Thomson et al. 1986; Holling and Meffe 1996; Stallman 2011).

4.6 Messages

4.6.1 Research Needs for Pricing Ecosystem Services in Agriculture

Research aims of pricing of ES in agriculture could put the following questions before formulation of the specific objective:

1. How to disaggregate the contribution of ESs to agricultural production?
2. How to scale the management practices governing the ESs?
3. How to value different offsets and trade-off of non-tangible ESs in agriculture?
4. How to design the targeted subsidies of inputs and/or incentive to smart agricultural practices that facilitates ESs?

The primary research needs are to quantify the ESs in agriculture with quantified data. The second important issue is to know, how native species in natural ecosystems provide services to enhance agricultural productivity and regulatory mechanisms. In this aspects, qualified data, evidence-based temporal information both at farm and landscape scale, and well-developed models supported by empirical data need to be generated. The third research needs are to develop proper understanding and conceptual framework into scales of ESs those provide benefit to human being. This is a prerequisite for pricing the services. Precisely, some of ESs are relevant and driven at farm scale (small scale, e.g. drainage, soil fertility, soil moisture retention, etc.) and some require the understanding the 'flow-of-services' in landscape scale (large scale; e.g. flow of pollination in natural system and agricultural crops; flow of water in upper catchment and down-command area of watershed, etc.) (Ricketts 2004; Zhang et al. 2007). The fourth research needs are to clearly demarcate the supply side and the beneficiaries of certain ES at watershed or landscape scale and how to put price on those services by clearly defining the offsets of those. For example, tree at catchment in hilly areas (supply side) facilitates groundwater recharging at downstream (command area) where the beneficiaries are completely different. In those cases, resources allocation and pricing of services need to be done at government or policy level by keeping the balance between human values (one setting) and supplier interest (another setting) for sustaining the proper ecosystem functioning. The fifth and critical researchable issue is to quantify the trade-offs within an agricultural landscape. Here, identifying the proper technologies for managing the trade-offs is important. For example, how much water resources, soil and native habitat should be exploited for getting desirable yield in an intensive agricultural system? The trade-off between private financial benefit and social losses from different management options for getting higher yield is a critical issue. For example, crop damage has to be restricted through pest control; but for this whether and what exact we should go for different management practices like maintaining of natural predator or labour-intensive spraying of chemical or 'drown-based' aerial spraying or only intercultural practices at farm levels are the researchable issues. The sixth research needs are to provide sufficient database to establish monetary value to ESs having wider market acceptance. Particularly, large number of research outputs in terms of original measured data and also model-based estimates are required to support the government/policymaker/private-public organizations to value the supporting and regulatory services having no-markets. In this aspect, genuine collaboration of economist and ecologist is desirable. These research output should link and validate with the real markets at varied locations. Economic trade-offs of services must be bridged with 'non-market' valuation method. The last but not the least, the research needs are an understanding of biophysical relationship among different provisional, regulatory, supporting and cultural ESs at landscape scale with the interdisciplinary approach by engaging political science, economics, ecology, hydrology and agriculture together (Zhang et al. 2007).

References

Adhikari K, Hartemink AE (2016) Linking soils to ecosystem services–a global review. Geoderma 262:101–111

Antle JM, Capalbo SM (2002) Agriculture as a managed ecosystem: policy implications. J Agric Resour Econ 1:1–5

Arnold JM, Campbell JG (1986) Collective management of hill forests in Nepal: the community forestry development project. In: Proceedings of the conference on common property resource management, vol 425. National Academy Press, Washington DC, USA, p 464

Bhattacharyya P, Bhaduri D, Adak T, Munda S, Satapathy BS, Dash PK, Padhy SR, Pattanayak A, Routray S, Chakraborti M, Baig MJ (2020) Characterization of rice straw from major cultivars for best alternative industrial uses to cutoff the menace of straw burning. Ind Crop Prod 143: 111919

Bhattacharyya P, Bisen J, Bhaduri D, Priyadarsini S, Munda S, Chakraborti M, Adak T, Panneerselvam P, Mukherjee AK, Swain SL, Dash PK (2021) Turn the wheel from waste to wealth: economic and environmental gain of sustainable rice straw management practices over field burning in reference to India. Sci Total Environ 775:145896

Bhattacharyya P, Neogi S, Roy KS, Dash PK, Nayak AK, Mohapatra T (2014) Tropical low land rice ecosystem is a net carbon sink. Agric Ecosyst Environ 189:127–135

Bhola K, Malhotra R (2014) Carbon credit capital: Indian commodity market 2020. Middle-East J Sci Res 22:1622–1629. https://doi.org/10.5829/idosi.mejsr.2014.22.11.21626

Bryan BA, Crossman ND (2013 Jun 1) Impact of multiple interacting financial incentives on land use change and the supply of ecosystem services. Ecosyst Serv 4:60–72

Climate change (2014) Synthesis report. In: Pachauri RK, Allen MR, Barros VR, Broome J, Cramer W, Christ R, Church JA, Clarke L, Dahe Q, Dasgupta P, Dubash NK (eds) Contribution of Working Groups I, II and III to the fifth assessment report of the Intergovernmental Panel on Climate Change

Duarte CM, Losada IJ, Hendriks IE, Mazarrasa I, Marbà N (2013 Nov) The role of coastal plant communities for climate change mitigation and adaptation. Nat Clim Chang 3(11):961–968

Dubey A, Lal R (2009 Oct 1) Carbon footprint and sustainability of agricultural production systems in Punjab, India, and Ohio, USA. J Crop Improv 23(4):332–350

Fourqurean JW, Duarte CM, Kennedy H, Marbà N, Holmer M, Mateo MA, Apostolaki ET, Kendrick GA, Krause-Jensen D, McGlathery KJ, Serrano O (2012 Jul) Seagrass ecosystems as a globally significant carbon stock. Nat Geosci 5(7):505–509

Gerber JS, Carlson KM, Makowski D, Mueller ND, Garcia de Cortazar-Atauri I, Havlík P, Herrero M, Launay M, O'Connell CS, Smith P, West PC (2016 Oct) Spatially explicit estimates of N2O emissions from croplands suggest climate mitigation opportunities from improved fertilizer management. Glob Chang Biol 22(10):3383–3394

GoI (2013) Pocket book on agricultural statistics 2013. Government of India, Ministry of Agriculture Department of Agriculture & Cooperation, New Delhi

Goldman RL, Thompson BH, Daily GC (2007) Institutional incentives for managing the landscape: inducing cooperation for the production of ecosystem services. Ecol Econ 64(2):333–343

Hillier J, Hawes C, Squire G, Hilton A, Wale S, Smith P (2009 May 1) The carbon footprints of food crop production. Int J Agric Sustain 7(2):107–118

Holling CS, Meffe GK (1996) Command and control and the pathology of natural resource management. Conserv Biol 10(2):328–337

IPCC (2006) National greenhouse gas inventories: land use, land use change and forestry. Institute of Global Environmental Strategies. Hayama, Japan

IPCC (ed) (2007) Agriculture. Cambridge University Press, Cambridge, United Kingdom/New York, NY, USA

IPCC (2014) Climate Change 2014: synthesis report. Contribution of Working Groups I, II and III to the fifth assessment report of the Intergovernmental Panel on Climate Change

IPCC (2019) Chapter 4: Forest land. In: Blain D, Agus F, Alfaro MA, Vreuls H (eds) Refinement to the 2006 IPCC guidelines for National Greenhouse gas Inventories (Vol. 4): agriculture, forestry and other land use (p. 68). IPCC (advance version)

IPCC (2021) In: Masson-Delmotte V, Zhai P, Pirani A, Connors SL, Péan C, Berger S, Caud N, Chen Y, Goldfarb L, Gomis MI, Huang M, Leitzell K, Lonnoy E, Matthews JBR, Maycock TK, Waterfield T, Yelekçi O, Yu R, Zhou B (eds) Climate change 2021: the physical science basis. Contribution of working group I to the sixth assessment report of the intergovernmental panel on climate change. Cambridge University Press, In Press

Ji C, Cheng K, Nayak D, Pan G (2018) Environmental and economic assessment of crop residue competitive utilization for biochar, briquette fuel and combined heat and power generation. J Clean Prod 192:916–923

Kritee K, Nair D, Zavala-Araiza D, Proville J, Rudek J, Adhya TK, Loecke T, Esteves T, Balireddygari S, Dava O, Ram K (2018 Sep 25) High nitrous oxide fluxes from rice indicate the need to manage water for both long-and short-term climate impacts. Proc Natl Acad Sci 115(39):9720–9725

Kumar P, Kumar S, Joshi L (2015) Socioeconomic and environmental implications of agricultural residue burning: a case study of Punjab. India, Springer Nature

Launio CC, Asis CA, Manalili RG, Javier EF (2016) Cost-effectiveness analysis of farmers' rice straw management practices considering CH_4 and N_2O emissions. J Environ Manag 183:245–252

Mandal B, Majumder B, Adhya TK, Bandyopadhyay PK, Gangopadhyay A, Sarkar D, Kundu MC, Choudhury SG, Hazra GC, Kundu S, Samantaray RN (2008 Sep) Potential of double-cropped rice ecology to conserve organic carbon under subtropical climate. Glob Chang Biol 14(9): 2139–2151

Melinda M (2013) F. Experiences with amine fluoride containing products in the management of dental hard tissue lesions focusing on Hungarian studies: a review

Muto Y, Yokokawa R (2022) Wetland Paddy fields as green infrastructure against flood. InGreen infrastructure and climate change adaptation. Springer, Singapore, pp 135–159

National Academy of Agricultural Sciences (2017) Innovative viable solution to rice residue burning in rice-wheat cropping system through concurrent use of super straw management system-fitted combines and turbo happy seeder. Policy Brief No. 2. National Academy of Agricultural Sciences, New Delhi

Nayak AK, Shahid M, Nayak AD, Dhal B, Moharana KC, Mondal B, Tripathi R, Mohapatra SD, Bhattacharyya P, Jambhulkar NN, Shukla AK (2019 Dec) Assessment of ecosystem services of rice farms in eastern India. Ecol Process 8(1):1–6

Paz-Alberto AM, Sigua GC (2013). Phytoremediation: a green technology to remove environmental pollutants

Rai R, Zhang Y, Paudel B, Acharya BK, Basnet L (2018 Sep) Land use and land cover dynamics and assessing the ecosystem service values in the trans-boundary Gandaki River basin, central himalayas. Sustainability 10(9):3052

Reed D, Washburn L, Rassweiler A, Miller R, Bell T, Harrer S (2016 Dec 13) Extreme warming challenges sentinel status of kelp forests as indicators of climate change. Nat Commun 7(1):1–7

Richmond A, Kaufmann RK, Myneni RB (2007) Valuing ecosystem services: a shadow price for net primary production. Ecol Econ 64(2):454–462

Ricketts TH (2004) Tropical forest fragments enhance pollinator activity in nearby coffee crops. Conserv Biol 18(5):1262–1271

Sandhu DS, Heinrich M (2005) The use of health foods, spices and other botanicals in the Sikh community in London. Phytother Res 7:633–642

Soam S, Kapoor M, Kumar R, Borjesson P, Gupta RP, Tuli DK (2016) Global warming potential and energy analysis of second generation ethanol production from rice straw in India. Appl Energy 184:353–364

Sreekumar A, Shastri Y, Wadekar P, Patil M, Lali A (2020) Life cycle assessment of ethanol production in a rice-straw-based biorefinery in India. Clean Techn Environ Policy 2:409–422

Stallman HR (2011) Ecosystem services in agriculture: determining suitability for provision by collective management. Ecol Econ 71:131–139

Suresh A, Reddy MP, Goud AS, Praneeth A, Manideep E, Sruti CL (2021). Conversion of crop residual waste in to fuel for thermal plants and fertilizer for farmers. Materials today: proceedings

Swain CK, Bhattacharyya P, Singh NR, Neogi S, Sahoo RK, Nayak AK, Zhang G, Leclerc MY (2016) Net ecosystem methane and carbon dioxide exchange in relation to heat and carbon balance in lowland tropical rice. Ecol Eng 95:364–374

Thompson RG, Singleton FD Jr, Thrall RM, Smith BA (1986) Comparative site evaluations for locating a high-energy physics lab in Texas. Interfaces 6:35–49

Van Mele P (2008) Biological control with the weaver ant Oecophylla longinoda in Africa: a review of research and development efforts to link farmers to organic markets. Outlooks on Pest Management 19(4):180

Wang J, Jiang L (2021) Intensive acorn processing in the early Holocene of southern China. The Holocene 14:09596836211041732

Wang L, Li F, Zhan Y, Zhu L (2016) Shifts in microbial community structure during in situ surfactant-enhanced bioremediation of polycyclic aromatic hydrocarbon-contaminated soil. Environ Sci Pollut Res 23(14):14451–14461

Wang H, Liu S, Mo J, Zhang T (2010) Soil-atmosphere exchange of greenhouse gases in subtropical plantations of indigenous tree species. Plant Soil 335(1):213–227

Watanabe I, Roger PR (1985) Workshop of wetland soils, characteristics, classification and utilisation, Los Banos. Philippines, International Rice Research Institute

Yan X, Cai Z, Ohara T, Akimoto H (2003) Methane emission from rice fields in mainland China: amount and seasonal and spatial distribution. J Geophys Res Atmos 108(D16)

Yan M, Luo T, Bian R, Cheng K, Pan G, Rees R (2015) A comparative study on carbon footprint of rice production between household and aggregated farms from Jiangxi, China. Environ Monit Assess 187(6):1–3

Yu X, Kai A, Gaodi X, Chunxia L (2011) Evaluation of ecosystem services provided by 10 typical rice paddies in China. Journal of Resources and Ecology 4:328–337

Zhang W, Ricketts TH, Kremen C, Carney K, Swinton SM (2007) Ecosystem services and dis-services to agriculture. Ecol Econ 64:253–260

5 Pricing of Renewable Energy-Based Applications in Agriculture

5.1 An Overview of Renewable Energy Applications in Agriculture

Renewable energy applications include solar energy, wind energy, biomass energy and hydrothermal energy. In agriculture, use of renewables is increasing at a faster rate as the agricultural activities are becoming energy intensive. It has been expected that energy use in agriculture in India needs to be increased from its present value 1.6 kW ha^{-1} to 2.5 kW ha^{-1} to meet the production target of next 20 years (NAAS 2018). In agricultural production system of present date, energy is required for fulfilling almost each activity starting from sowing to harvest at the end, e.g. land preparation using tillage implements and tools, sowing of seeds using mechanized seed drill, intercultural operations (e.g. tractor drawn weeding, hoeing, spraying of pesticides, etc.), harvesting using combined harvester, etc. Even, the post-harvesting processing, e.g. threshing, winnowing, drying of produces, value addition, etc., needs high amount of energy. A detailed account of energy use in agriculture is given in Elsoragaby et al. (2019) and specifically in arid agriculture is reported by Singh et al. (2002, 2003 and 2004). In India, about 18% of total energy consumption of the country is used in agriculture sector. Most of the needs of energy in agriculture section is conventionally met through fossil fuel-based sources, e.g. diesel and coal based thermal power plants. Since fossil fuels are not the infinite sources and emit large amount of CO_2 gasses in atmosphere during its utilization, there is need to adopt alternative sources of energy. Renewable energy plays a big role to meet the large share of energy need in agriculture in future to develop sustainable and clean agricultural production system. At present, the most common use of renewable energy in agriculture are solar photovoltaic (PV) pump-based irrigation and solar thermal devices (e.g. solar cooker, solar drier) for post-harvest processing. However, efforts are ongoing to apply renewable sources in other activities too, e.g. solar powered tractors and machineries, utilization of agro-wastes for generation of energy, aero generators for utilization of wind energy in agriculture, etc.

P. Bhattacharyya et al., *Pricing of Ecosystem Services in Agriculture: A Basis of Crop Insurance*, https://doi.org/10.1007/978-981-19-4416-1_5

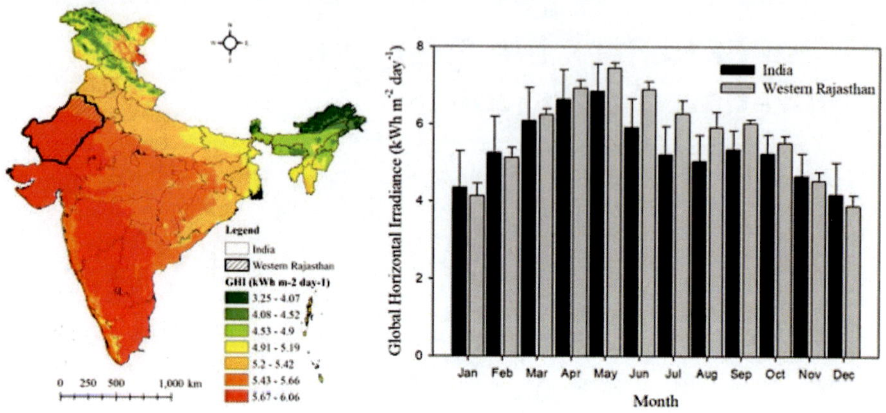

Fig. 5.1 Availability of global horizontal irradiance (GHI) in India and its monthly distribution

Total estimated renewable energy potential in India is about 900 GW, among which wind power potential is about 102 GW at 80 m mast height above ground level (agl), solar power potential is about 750 MW, and bioenergy potential is about 25 GW. The average solar irradiance on horizontal surface in India is 5.6 kWh m^{-2} day^{-1}, whereas in Leh and Ladakh it is highest (7–7.5 kWh m^{-2} day^{-1}) followed by in western India (6–6.5 kWh m^{-2} day^{-1}) (Fig. 5.1) (Santra 2015, 2019). Rajasthan shares the maximum amount of potential solar energy in India with a potential of about 142 GW. This is mainly due to incidence of high solar irradiance in western Rajasthan throughout the year (~6.5–7.5 kWh m^{-2} day^{-1} during summer to 4.5–5.5 kWh m^{-2} day^{-1} during winter) and high occurrence of cloud free day in a year (>300 days in a year). Available solar irradiation and utilizable energy for any location in India can also be viewed from http://mnre.gov.in/sec/solar-assmnt.htm.

Total wind power potential of the country is about 102 GW at 50 m height (Fig. 5.2), out of which onshore potential is 49,130 MW. Most of the wind power potential lies at the state of Gujarat, Andhra Pradesh, Tamil Nadu and Karnataka with an estimated potential of 35.07, 14.49, 14.15 and 13.59 GW, respectively. In deserts of India covering western Rajasthan, the wind energy potential is about 5.05 GW. Wind power density at 50 m height above ground level (agl) in western Rajasthan is about 200–250 W m^{-2}. Considering the availability of high-speed wind corresponding to rated wind speed of a turbine for a period of 6 hours per day, wind energy of 0.36–0.45 kWh m^{-2} day^{-1} can be harnessed in western Rajasthan considering a turbine efficiency of 30%.

Total amount of biomass available in India is estimated about 500 million metric tons per year. It has also been estimated that surplus biomass availability is about 120–150 million metric tons per annum covering agricultural and forestry residues corresponding to a potential biomass energy generation of about 17,536 MW. Highest biomass generation potential is observed in Punjab with a total potential of about 3172 MW. Apart from it, about 5000 MW additional power could be

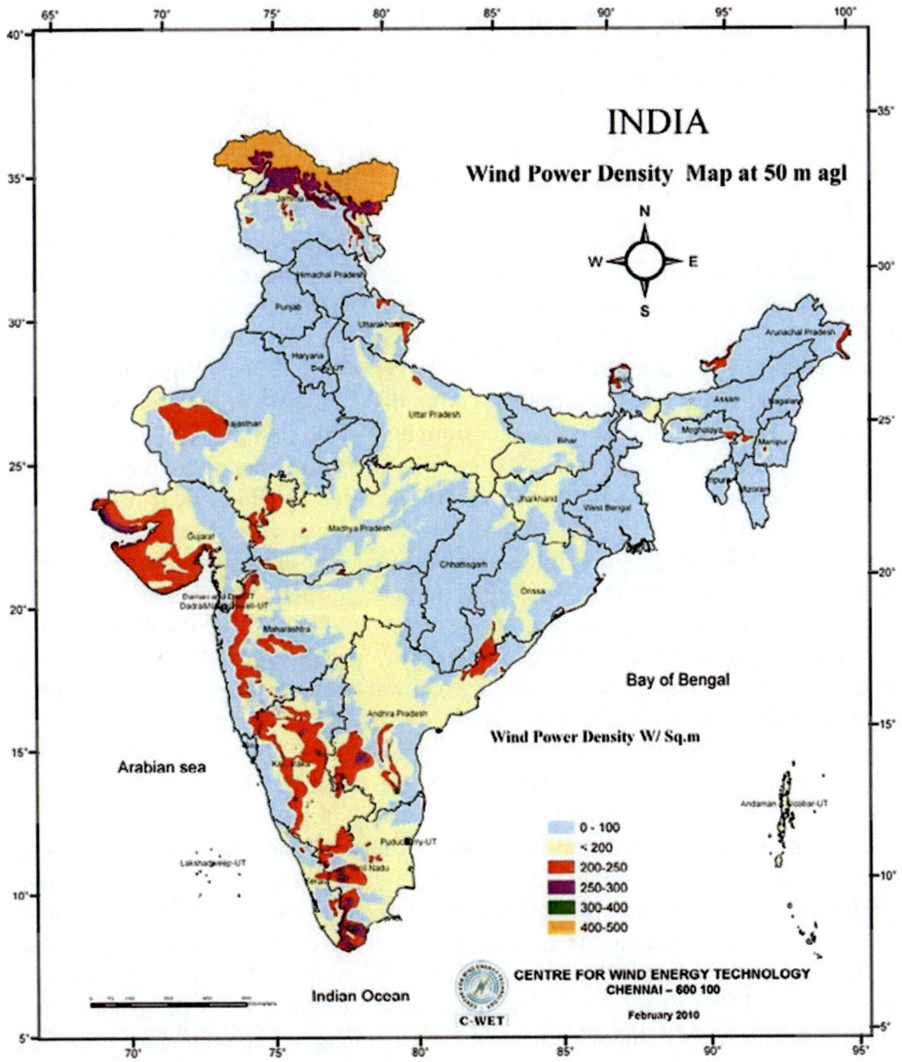

Fig. 5.2 Wind power potential map of India

generated through bagasse-based cogeneration in the country's 550 Sugar mills. Most of these bagasse-based power generation potential of the country lies in Maharashtra and Uttar Pradesh with a potential of 1250 MW from each state. Waste from agricultural farm, e.g. cow dung, can also be converted to energy and there is a potential of 2554 MW across the country with Maharashtra at the top of the list.

In world, each country has shifted their focus from fossil fuel-based power generation to renewable. Total renewable energy installation capacity in the world has been increased from 1329 GW during 2011 to 2799 GW during 2020 (IRENA

2021). As a result, about 28% of global electricity generation is now met from renewable sources. China ranks in the top in renewable energy-based electricity generation in the world with a generation of 288 TWh (terra watt-hour), whereas it is followed by the Unites States, European Union (EU) and India (IEA 2021). Wind energy contributes (275 TWh) highest in the total renewable based electricity generation followed by solar (145 TWh), hydrothermal (140 TWh) and bioenergy (72 TWh).

In India, the national solar mission has been in progress since 2009 and targets were fixed in three phases with a total renewable installation of 175 GW in the country by 2022. Among these, the target for solar PV, wind energy, biomass energy and other renewables were 100 GW, 60 GW, 10 GW and 5 GW, respectively. This target has further been revised with a new target of 450 GW by 2030. It has been envisaged that following these targets, about 40% and 60% of total electric power generation capacity in India will be met from renewable sources by 2022 and 2030, respectively. Further, Govt of India has committed a 100% renewable power system target by 2050 and net zero carbon emission target by 2070, declared in COP 26 climate meeting held in Glasgow during November, 2021. Agriculture sector has great scope in contributing these national targets by utilizing renewable energy in different applications (e.g. solar PV pumping system for irrigation) and by generating energy from agricultural land by installations of solar PV system, wind turbines and establishing agro-waste based power generation. After realizing the potential of agriculture sector in meeting renewable energy targets of the country, Govt of India has launched the scheme '*Kisan Urja Suraksha evam Utthan Mahabhiyan*' (KUSUM) in 2019, under which solar PV systems and solar PV pumps are targeted to install in farmers field, details of which is available in https://mnre.gov.in/ and https://www.india.gov.in/spotlight.

During last one-decade, total renewable installation capacity in India has been increased from 14.4 GW at the beginning of 2009 to 95.8 GW by the end of March, 2021 (https://mnre.gov.in/). In India, solar power installations dominate in total renewable energy installations accounting for 43% of total installations and it includes ground mounted PV system (35.65 GW), roof top PV system (4.44 GW) and off-grid SPV system (1.15 GW). Wind energy installations is the second most important contributor in India's renewable energy industry, accounting for over 41% of installed capacity (39.25 GW). It is followed by biomass-based power generation (10.31 GW) and small hydro power (4.77 GW). Rajasthan shares ~21% of the total solar power installed capacity in the country, whereas Karnataka and Gujarat share about 15% and 13%, respectively. Tamil Nādu and Gujarat dominate the total wind installation in India by sharing 47.1% of total installed capacity. Among off-grid PV installations, about 3,38,701 solar pumps have been installed across the country by the end of February 2022.

5.2 Ecosystem Services of Solar Energy-Based Applications

Ecosystem services of renewable energy systems and devices are calculated using their provisioning services, regulating services and supporting services. Here, cultural services are ignored since it is very difficult to quantify the non-material benefit gained by people after adopting the renewable energy-based system. Provisioning services are defined as actual benefit obtained by people after adopting a technology, which may be a specific agricultural production system or eco-friendly agricultural implements, tools and devices. These benefits may be food, fuel, fodder, water, etc. If renewable energy devices or systems is used by people, the provisioning services (PS) include the produced equivalent energy by the system or device, which may be further quantified in terms of money using the price value of energy.

$$\text{PS (Rs)} = \text{Energy equivalent (kWh)} \times \text{unit price of energy} \left(\text{Rs kWh}^{-1}\right) \quad (5.1)$$

However, the price value of provisioning service needs to be subtracted from the cost involved to establish the system as well as the cost involved in maintenance and repair of the system till its life cycle. Thus, the net price value of provisioning service can be obtained. To calculate the benefit and cost involved in a renewable energy-based device or system during its whole life cycle period, life cycle cost (LCC) approach needs to be applied.

In the LCC approach, total life cycle cost of a system or device is calculated by combining capital cost, maintenance cost, replacement costs for damagaed components and operational cost. Similarly, all future benefits (B) of the device in its whole life cycle need to be combined together to obtain total life cycle benefit (LCB). While calculating these costs and benefits, all future costs (C) and benefits (B) are converted to present worth (PW) after considering the relative rate of inflation (i) and discount rate (d) as follows:

$$\text{PWC}_{jk} = C_k \times \left[\frac{(1+i)}{(1+d)}\right]^j \quad (5.2)$$

$$\text{PWB}_{jk} = B_k \times \left[\frac{(1+i)}{(1+d)}\right]^j \quad (5.3)$$

where PWC_{jk} is the present worth of kth component of the cost in j^{th} year, PWB_{jk} is the present worth of kth component of benefit in j^{th} year, k is the number of component in cost or benefit ($k = 1,2,3,\ldots m$), m is the number of components in cost or benefit, j is the time period in years ($j = 1,2,3\ldots n$), n is the life cycle period in years. In conventional economic analysis, cost component is mentioned as cash outflow, whereas benefit component is referred as cash inflow. Net benefit of the system is referred as net cash flow. Relative rate of inflation is the relative increase or decrease in prices of a commodity as compared to inflation rate. In case, the the price of any commodity is expected to increase by following the actual inflation rate, then relative rate of inflation is considered zero. Discount rate is defined as the real value of money in future. In most developing economies of the world it is about 8–12%, and therefore 10% is considered as an average value for most economic analyis. A discount rate of 10% per year indicates that if present value of a commodity is Rs 100 in current year, then in the next year its value will be Rs 110. Conversely, if the future value of any commodity is expected to be Rs 100, then at the present year its real value will be Rs 90. In case, the relative rate of inflation is considered zero, the Eqs. (5.2) and (5.3) may be rewritten as follows:

$$\mathrm{PWC}_{jk} = \frac{C_k}{(1+d)^j} \tag{5.4}$$

$$\mathrm{PWB}_{jk} = \frac{B_k}{(1+d)^j} \tag{5.5}$$

For a future single cost in j^{th} year, the present worth of that cost is calculated using Eq. (5.4), whereas for single future cash inflow in j^{th} year, the present worth of that cash inflow is calculated using Eq. (5.5). For future multiple cash outflows and inflows, these are to be converted to present worth for each component and each year and then needs to be cumulated to obtain total cash outflow and cash inflow as follows:

$$\mathrm{PWC}_j = \sum_{k=1}^{m} \mathrm{PWC}_{jk} = \sum_{k=1}^{m} \frac{C_k}{(1+d)^j} \tag{5.6}$$

$$\mathrm{PWB}_j = \sum_{k=1}^{m} \mathrm{PWB}_{jk} = \sum_{k=1}^{m} \frac{B_k}{(1+d)^j} \tag{5.7}$$

Net present value (NPV) of the system at the end its life cycle period is calculated as follows:

$$\mathrm{NPV} = \sum_{j=1}^{n} \mathrm{PWB}_j - \sum_{j=1}^{n} \mathrm{PWC}_j \tag{5.8}$$

Following Eq. (5.8), NPV for j^{th} year ($j = 1,2,3\ldots n$) within the life cycle period may be calculated. At initial years, NPV is generally found negative and increases in subsequent years. For calculation of annualized life cycle cost (ALCC), annuity factor (AF) is calculated for a period of life cycle as follows:

$$\mathrm{AF} = \frac{\frac{1+i}{1+d} - 1}{\frac{1+i}{1+d}\left[\left(\frac{1+i}{1+d}\right)^n - 1\right]} \tag{5.9}$$

$$\mathrm{ALCC} = \mathrm{LCC} \times \mathrm{AF} \tag{5.10}$$

In addition to NPV and ALCC, payback period is most often calculated. Payback period of a system is the time required for capital investment to be repaid by the cash outflow generated by the system. The simple payback period is calculated by dividing the invested amount with the annual cash inflow. However, discounted payback period is calculated as the time period when cumulated present value of cash inflow equals to cumulated present value of cash outflow. This time period is often referred as breakeven period. The payback period is expressed in years and fractions of years. Shorter is the payback period, lower is the risk for investment. In the payback analysis, ALCC as shown in Eq. 5.9 can be considered as cash outflow for a year. The cash inflow is calculated by adding the income generated by the farmer and the cost of energy saved by using the solar device.

The payback period (N) can also be calculated by using mathematical formula as follows (Nahar 2001):

$$N = \frac{\log\left[\frac{E-M}{a-b}\right] - \log\left[\frac{E-M}{a-b} - C\right]}{\log\left[\frac{1+a}{1+b}\right]} \tag{5.11}$$

where E is the price value of energy produced by the system per year (Rs), M is maintenance cost per annum (Rs), C is cost of the device (Rs), a is the compound interest rate per annum and b is inflation rate in energy and maintenance per annum.

Internal rate of return (IRR) is another economic metric to calculate the profitability of any investment. This is calculated by following the same methodology as used in NPV calculation. The IRR is defined as the discount rate at which NPV equals to zero at particular time period within its life cycle period and is calculated as follows:

$$\mathrm{NPV} = 0 = \sum_{j=1}^{n}\sum_{k=1}^{m}\frac{B_k}{(1+\mathrm{IRR})^j} - \sum_{j=1}^{n}\sum_{k=1}^{m}\frac{C_k}{(1+\mathrm{IRR})^j} \tag{5.12}$$

The regulating services of the renewable energy devices are quantified through its contribution in reduced carbon footprint. It is expected that any renewable energy devices or system will reduce the CO_2 emission in atmosphere, which is otherwise released in atmosphere due to use of conventionally used fuel-based system. Carbon footprint is defined as the equivalent amount of CO_2 emission per unit of energy produced by the system. The carbon footprint of any energy generation system or device can be calculated from the emission factor (EF) of the method of energy generation in the system after dividing it with their efficiency.

$$\text{Carbon footprint}\left(CO_2\text{-eq kWh}^{-1}\right) = \frac{\text{Emission factor (EF)}\left(CO_2\text{-eq kWh}^{-1}\right)}{\text{Efficiency of energy generation}} \tag{5.13}$$

The EF of coal-based energy production system is about 0.82 kg CO_2-eq kWh^{-1} (Central Electricity Authority 2018). Based on the net calorific value of diesel, EF of diesel fuel is 72.6 g CO_2-eq. MJ^{-1}, which is equivalent to 0.26 kg CO_2-eq kWh^{-1}. Otherwise, the EF of diesel fuel can also be calculated from its volumetric based emission factor of 10.21 kg CO_2-eq $gallon^{-1}$ or 2.697 kg CO_2-eq l^{-1} and the energy value of diesel as 10.5 kWh l^{-1}. In case of solar PV based electricity production system, the EF is reported as 6.15 g CO_2-eq kWh^{-1} during its operation and maintenance cycle (Nugent & Sovacool 2014). However, if total life cycle of a PV system starting from fabrication to utilization is considered, EF will be slightly higher, 49.9 g CO_2-eq kWh^{-1}. After calculating the carbon footprint (CF), it is converted to monetary value of regulating service (RS) by multiplying it with social cost of carbon (SCC) to obtain the final value of regulating services (RS) as follows:

$$\text{RS}\ (\$\ \text{y}^{-1}) = \text{CF}\ (\text{t CO}_2\text{–eq kWh}^{-1}) \times \text{Energy generation}\ (\text{kWh y}^{-1}) \times \text{SCC}\ (\$\ \text{t CO}_2\text{–eq}^{-1}) \quad (5.14)$$

The value of SCC for India is reported as highest across globe and it is US \$86 t^{-1} CO_2-eq, followed by the United States, where SCC is US \$48 t^{-1} of CO_2-eq (Ricke et al. 2018).

Adoption of renewable energy devices saves the involvement of labours during operation and maintenance of the system, which is otherwise required with conventional fuel based or grid connected system. Such savings in man-days can be converted to monetary value to calculate the supporting services (SS) of renewable energy devices as follows:

$$\text{SS}\ (\$\ \text{y}^{-1}) = \text{Savings in Man-days}\ (\text{y}^{-1}) \times \text{Cost of man-days}\ (\$\ \text{man-day}^{-1}) \quad (5.15)$$

The average cost of man-days in India as per the Mahatma Gandhi National Rural Employment Guarantee Act (MGNREGA) of Govt of India is about \$3.72 $man\text{-}day^{-1}$ (INR 282 $man\text{-}day^{-1}$ and \$1 = INR 75.84).

In few cases, adoption of renewable energy system or device helps in harvesting of rainwater followed by recycling it in the system for its utilization, which otherwise needs to be extracted from deep groundwater resources. Such savings in water use by replacing the groundwater withdrawal can be calculated to monetary value to quantify the supporting services (SS) as follows:

$$\text{E}\ (\text{kWh ha-mm}^{-1}) = \rho\ (\text{kg m}^{-3}) \times v\ (\text{m}^3) \times 10 \times g\ (\text{m s}^{-2}) \times h\ (\text{m}) \times 3600 \quad (5.16)$$

$$\text{SS}\ (\text{ha-mm}^{-1}) = \text{E}\ (\text{kWh ha-mm}^{-1}) \times \text{Energy tarif rate}\ (\text{kWh}^{-1}) \quad (5.17)$$

where E is the energy involved in withdrawal of groundwater, ρ is the density of water, v is the volume of withdrawal groundwater, g is the acceleration due to gravity (9.8 m s^{-2}) and h is the depth of groundwater. Apart from savings in groundwater withdrawal, few renewable energy based system (e.g. ground mounted PV system in agricultural field) conserves soil moisture and thus reduces the amount of irrigation water (ha-mm), which further can be converted to SS by the following Eq. (5.17).

Another supporting services of renewable energy system is the reduction in loss of nutrient-rich soil particles from agricultural field since it provides ground cover and reduces the causative factors of erosion. Such loss in soil and associated loss of nutrient can be quantified to monetary value to account the supporting services as follows:

$$\mathrm{SS}\left(\mathrm{ha}^{-1}\,\mathrm{y}^{-1}\right) = \mathrm{SL}\left(\mathrm{t\,ha}^{-1}\,\mathrm{y}^{-1}\right) \times \mathrm{NC}\left(\mathrm{kg\,t}^{-1}\right) \times \mathrm{CN}\left(\mathrm{kg}^{-1}\right) \quad (5.18)$$

where SL is reduction in soil loss rate, NC is the nutrient content of eroded soil and CN is the cost of nutrient. For example, the cost of nutrient-based subsidy rates of Govt of India during 2021–2022 for *N*, *P*, *K* and *S* are about \$0.25 kg^{-1} (INR 19 kg^{-1}), \$0.60 kg^{-1} (INR 45 kg^{-1}), \$0.13 kg^{-1} (INR 10 kg^{-1}) and \$0.03 kg^{-1} (INR 2 kg^{-1}), respectively. (https://pib.gov.in/PressReleaseIframePage.aspx?PRID=1763349) (\$1 = INR 75.48).

In the following sections, ecosystem services of different renewable energy systems and devices are discussed. Renewable energy systems relevant for agriculture sector are categorized into four major groups: solar thermal devices, solar PV based applications and devices, wind energy based system and biomass-based power generation.

5.2.1 Solar Thermal Devices

The principle of solar thermal technology is the generation of heat by trapping the incident solar irradiation inside an insulated enclosure chamber through a transparent glass collector and further using the generated heat energy for different purposes, e.g. drying, heating, cooking. Based on the principle, several types of solar thermal devices have been developed for post-harvest processing and value addition of agricultural produces (Pande et al. 1980, 2009; Rao et al. 2017; Singh et al. 2020). For example, inclined solar drier, animal feed solar cooker, solar water heater, solar distillation unit, etc. have great potential in improving the livelihood farmers in remote villages (Fig. 5.3).

5.2.1.1 Solar Dryer

Solar dryer is most used solar thermal device to dehydrate fruit, vegetables and grains efficiently (Pande and Thanvi 1988; Poonia et al. 2017a) (Fig. 5.3a). As compared to oven sun drying, solar dried products are free from contamination of dust and other foreign pathogens and thus are always preferred to obtain higher market price. Moreover, there is less chance of insect infestation and spoilage of products in solar drying process due to protection of sudden rain events. Among these forced and natural circulation type of driers, inclined solar driers are most used. Other types are solar cabinet drier, solar cooker cum drier, phase change material based drier, forced convection type drier with solar air heaters and electric blowers,

Fig. 5.3 Different solar thermal devices having potential in agriculture

etc. A drier of 1 m^2 glass collector can dry 10–12 kg of fruits and vegetables in 12–24 h, depending upon the initial moisture content of the produces. Depending on the cost of solar dried products, which generally varies as per fruits and vegetables, provisioning services can be calculated. For example, the market value of solar dried date palm in India is about \$4.44 kg^{-1} (INR 335 kg^{-1}), whereas for fresh date fruit the market price varies from \$1.06 kg^{-1} to \$1.59 kg^{-1} (\$1 = INR 75.48). Moisture content in dried date fruit is about 25%, whereas for fresh date fruit it is about 60%. Therefore, 1 kg of fresh date fruit produces about 0.533 kg of dried date fruit. Drying of date palm fruit through solar drier thus may provide an additional income of \$12.48 considering the loading capacity of 1 m^2 solar drier is 12 kg fresh fruit of date palm. Thus, the provisioning services of a solar drier of 1 m^2 collector area is about \$12.48 for a single drying. Considering the average fruit yield of 40 kg per tree per year planted at 10 m × 10 m spacing, 4 t of fresh date fruit is expected from 1 ha orchard per year. Total 20 m^2 solar drier is calculated to be sufficient to dry 4 t of fresh date palm fruit from 1 ha orchard in about 16 batches, whereas each batch of drying takes about 2–3 days. Thus, the provisioning service of solar drier is about \$4163 ha^{-1} y^{-1}. Considering the capital cost of 1 m^2 is about \$119 per m^2 collector area, the capital investment for 20 m^2 solar drier is about \$2380. The solar drier can be used for drying onion, okra, carrot, garlic, tomato, chilli, ber, date, spinach, coriander, salt coated aonla, etc. Typically, the thermal efficiency of the dryer is 17.57%. Apart from the economic benefit, solar drier can save about 290–300 kWh m^{-2} y^{-1} equivalent energy, which is otherwise used in conventional oven drying. Thus, solar drier of 1 m^2 collector area can provide a support service of $28.82–29.81 per year considering an electricity tariff rate of $0.1 kWh^{-1}. Use of solar drier can also reduce 1127 kg of CO_2 y^{-1} (Pande et al. 2016), if replaces the conventional electricity operated oven drier and thus again provides a regulating service of about $96.92 per year considering the social carbon credit value of $86 per t CO_2-eq in India ($1 = INR 75.48).

5.2.1.2 Animal Feed Solar Cooker

Animal feed solar cooker, which is generally used for boiling of animal feed is another solar thermal device, mostly used by farmers (Nahar 1994; Nahar et al. 1996; Poonia et al. 2017b) (Fig. 5.3b). Conventionally, the feed for cattle is boiled through burning of fuel wood, cattle dung cake and agricultural wastes before feeding to enhance quantity and quality of milk. Animal feed solar cooker can be successfully used to boil different types of animal feed, e.g. crushed barley (*Jau Ghat*), cluster bean split (guar korma), cluster bean powder, cotton seed and gram churi during daytime within 4–5 h. Use of animal feed solar cooker saves time of women to prepare the feed, which can be utilized for another domestic work. The efficiency of the animal feed solar cooker is about 21.8%. The animal feed solar cooker saves about 1059 kg of fuel wood per year, which is equivalent to 3611 MJ (~1003 kWh) of energy. In case, liquified petroleum gas (LPG) is used for cooking animal feed, the equivalent amount of LPG saving per year is 78.33 kg, considering the specific calorific value of LPG as 46.1 MJ kg^{-1}. Saving of 78.33 kg of LPG per year by adopting animal feed solar cooker can provide a support service of $65 per year (LPG cost is $0.83 kg^{-1}). Considering the electricity tariff rate of $0.1 per kWh of electric energy, the device can save $100 per year if the electricity operated ovens are used for the purpose. Moreover, the use of the solar cooker for animal feed would result on the reduction in emission of 1442.64 kg of CO_2 per year (Pande et al. 2016). Reducing CO_2 emission indirectly contributes to social carbon credits @ $86 per ton of CO_2-eq. Thus, animal feed solar cooker provides the regulating services of $124 per year ($1 = INR 75.48).

5.2.1.3 Solar Cooker

Solar cooker is another thermal device, which is used for cooking purpose and can replace cooking gas or liquefied petroleum gas (LPG), firewood, agricultural waste and cow dung cake (Nahar 1998) (Fig. 5.3c). Different types of solar cooker is available, e.g. solar oven, hot box solar cooker, non-tracking solar cooker. Among these, non-tracking solar cooker does not need frequent tracking of cooker towards sun. Solar cookers are generally used for boiling rice, vegetables, pulses, etc. The overall efficiency of the solar cooker is 24.6%. Use of the non-tracking solar cooker can save about 30–40% of fuel requirement. The efficiency of the non-tracking solar cooker is 29.5%. The energy for cooking per person is about 900 kJ of fuel equivalent per meal. The large size non-tracking solar cooker can cook for about 10 persons, and it will save 50% of cooking fuel per meal. Therefore, it can save 9 MJ of energy per meal and 2587.5 MJ (~719 kWh) of fuel equivalent per year. Considering an electricity tariff rate of \$0.1 kWh^{-1}, non-tracking solar cooker saves about \$71.9 per year, if the solar cooker replaces electric operated oven for cooking purpose. In case the cooker replaces LPG cylinder for cooking purpose, it saves about 56.13 kg of LPG per year, which is equivalent to about \$46.59 per year (LPG cost is \$0.83 kg^{-1}). Similarly, solar cooker provides the regulating services by replacing the LPG or electricity dependent cooking, which otherwise contributes to CO_2 emission in atmosphere (Pande et al. 2016).

5.2.2 Solar Photovoltaic Applications

5.2.2.1 Solar PV Pumping System

Solar photovoltaic (PV) pumps are the most prudent solution to provide assured power supply in agricultural field for irrigation purpose (Pande et al. 2003; Kumar et al. 2015; Santra et al. 2017a). Typically, a solar PV system consists of: (i) PV modules, (ii) mounting structure, (iii) pump unit (AC/DC) and (iv) tracking system (Santra et al. 2016). The capacity of solar PV pumping system depends on several factors, e.g. depth of water table for pumping (suction head), delivery pressure head, available solar irradiation and its diurnal variation, irrigation demand. If the target is to irrigate a small catchment of about 1 acre from small capacity surface water reservoir or farm pond, then 1 HP solar PV pumping is sufficient (Fig. 5.4) (Santra 2021). If the solar PV pump is to be used for drawing more deep groundwater from wells or tube wells, then the capacity may be 3 HP or 5 HP and is mostly used by farmers.

The capital cost of a solar PV pumping system is contributed by PV module, pumping unit, inverter, mounting structure and accessories/cables. Life cycle costs for a 3 HP and 5 HP solar PV pumping systems are presented in Table 5.1 (Santra

Fig. 5.4 Solar photovoltaic pumps at experimental field of ICAR-CAZRI, Jodhpur, (**a**) 1 HP AC pumping system with 1400 W_p PV modules and (**b**) 1 HP DC pumping system with 900 W_p PV modules

Table 5.1 Life cycle cost of solar PV pumping system for irrigation

Sr. No.	Parameters	Solar PV pumping system	
		3 HP system	5 HP system
1.	Life cycle	25 years	25 years
2.	PV panel cost ($0.53/$W_p$)	$1855	$3127
	AC pump cost	$397	$464
	Inverter cost	$265	$331
	Mounting structure	$331	$397
	Cables and accessories	$133	$133
	Miscellaneous cost (30% of total cost)	$894	$1336
	Total capital cost	$3872	$5787
3.	Lifetime maintenance cost (1% of the capital cost)	$352	$525
4.	Replacement cost of AC pump (at eighth year and 16th year)	$288	$320
5.	Replacement cost of inverter (at tenth year and 20th year)	$143	$179
6.	Total life cycle cost	$4654	$6811
7.	Annualized life cycle cost	$513 y^{-1}	$750 y^{-1}

et al. 2016). Total life cycle cost for 3 HP solar PV pumping system was found $4654, whereas it was $6811 for 5 HP system. Similarly, the life cycle cost of 1 HP (AC) solar PV pumping system is about $2553 (Santra 2021) ($1 = INR 75.48). The cash inflow of a solar PV pumping system is the energy it provides for pumping purpose.

The cash inflow after adoption of solar PV pumping system is the price value of energy for pumping, which is otherwise used for meeting the cost for diesel and electricity in conventional diesel operated and grid connected pumps. If we consider average use of diesel pump in a year for irrigation is 60 days and 6 hrs per day the cost of diesel per year for a 3HP diesel pumping set is about $179, considering (i) diesel price @ $0.75 l^{-1}, (ii) energy value of diesel is 10.5 kWh l^{-1}, (iii) efficiency of diesel pump is 30–35% and (iv) energy generation by diesel pump is 3.4 kWh l^{-1}. Thus, a farmer can save $179 y^{-1} if a farmer uses solar pump by replacing a diesel pump, which is accounted as provisioning services of solar PV pumping system. The annual cash outflow for a 3 HP solar pump system is $513 as mentioned in Table 5.1. Therefore, the payback period of a 3 HP solar pump will be 2.9 years. Similarly, if the solar pump replaces grid connected electric pumps, it saves electricity charge of $80.5 in a year considering the similar duration of its use as mentioned above and electricity tariff of $0.1 kWh^{-1} and thus the payback period will be 9.6 years. Therefore, it is seen that replacing diesel operated pumps by solar pump is highly beneficial to a farmer. Ensuring energy security for irrigating the crops at right time is expected to improve the crop productivity, which may be about 10–15% increment and thus contributes in net cash inflow and can be accounted as the supporting services of solar PV pumping system. Solar PV pumping system has additional advantages over other pumping systems, e.g. (i) PV panels of a solar

Table 5.2 Carbon footprint of solar PV pumping system, grid connected pumps and diesel operated pumps for irrigation purpose

Sr. No	Types of irrigation pumps	Carbon footprint (kg CO_2-eq kWh^{-1})	Carbon footprint (kg CO_2-eq ha-mm^{-1a})
1.	Solar PV pump	0.011	0.009
2.	Grid connected electric pump	1.439	1.214
3.	Diesel operated pump	0.743	0.382

[a]Pump capacity = 1 HP; Total head = 20 m

pumping system reduce the CO_2 emission in atmosphere; (ii) during off time, electricity generated by the solar PV pumping system may be used for domestic needs and for operating small farm machines; (iii) solar PV pumping system may be used in far remote locations, where electric grids are not available.

The carbon footprint of 1 HP solar PV pumping system along with diesel and grid connected pumping system is given in Table 5.2. For solar PV pumping system carbon footprint is almost negligible as compared to grid connected and diesel operated pumps. Grid connected pumps have highest carbon footprint, which is mainly because of use of coal as the major input in thermal power-based grid networks of the country and the carbon emission factor of coal is higher than diesel fuel.

The carbon footprint of irrigation pump can also be converted to equivalent carbon footprint per ha-mm irrigation depth following Eq. (5.16). Such conversion revealed that the carbon footprint per ha-mm irrigation is 1.214 kg CO_2-eq for grid connected electricity pump, whereas for diesel operated pumping system it is about 0.382 kg CO_2-eq. The reduction in carbon footprint after adoption of solar PV pumping system by replacing grid connected and diesel pump is its regulating service. Depending on the total amount of irrigation water applied for crops grown under solar PV pumping system, the price value of regulating service of solar PV pumping system can be calculated. For example, water requirement of cumin crop in arid region of Rajasthan is about 375 ha-mm. Thus, the reduction in carbon footprint is about 455 kg CO_2-eq ha^{-1} if solar PV pump replaces grid connected pump and is about 143 kg CO_2-eq ha^{-1} if solar PV pump replaces diesel pump. Thus, the regulating service of solar PV pumping is about \$39.13 and \$10.79, respectively, for replacing grid connected pump and diesel pump (SCC is \$86 t^{-1} CO_2-eq and \$1 = INR 75.48).

5.2.2.2 Solar PV Sprayer/Duster

Approximately, 35% of the crop production is damaged if pest and diseases are not controlled at right time. Uniform spraying of liquid formulations or dusting of plant protection chemicals throughout the crop field is very important for effective control of pest and diseases. Keeping in mind these requirements, several solar PV operated equipment have been designed and developed, e.g. solar PV sprayer, solar PV duster.

Fig. 5.5 Solar PV sprayer

Table 5.3 Cost of different components of developed solar PV sprayer

Sr No.	Component	Cost
1	Solar panels	\$80
2	DC motor	\$73
3	Batteries	\$20
4	Charge controller	\$16
5	Trolley	\$126
6	Pesticide tank	\$3
7	Accessories (nozzle, pipe, plywood, GI sheet, paints, etc.)	\$46
	Total	\$364

Solar PV sprayer is used for spraying of agricultural chemicals in agricultural field (Swami et al. 2016; Santra et al. 2019) (Fig. 5.5). To provide energy to DC pump (60 W) of the PV sprayer, 120 W_p capacity (60 W_p × 2 Nos) solar PV modules are connected so that the produced energy may be directly used by DC motor.

The detailed cost of different components of the developed sprayer is presented in Table 5.3. The system is designed in such a way that its use will be multipurpose. It provides additional facility to farmers for fulfilment of domestic use of energy.

Solar PV duster is used for application of dust formulation pesticides, e.g. sulphur dust, malathion powder (Santra et al. 2019; Singh et al. 2020) (Fig. 5.6). It essentially comprises of a PV module (7.5 W_p), a metal carrier, storage battery (12 V, 7 Ah) and especially designed compatible dusting unit.

5.2.3 Agrivoltaic System

Agrivoltaic system (AVS) integrates PV generation and agricultural production in a single system keeping in view the requirement of both for civilization to progress and limitation of arable lands to meet future food production target (Santra et al. 2017b). The concept of AVS and its relevance is schematically presented in Fig. 5.7.

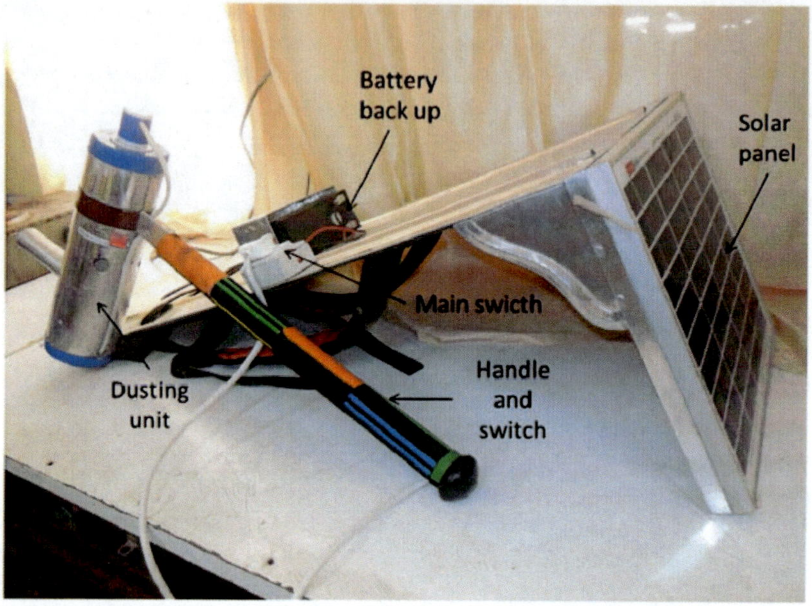

Fig. 5.6 Solar PV duster

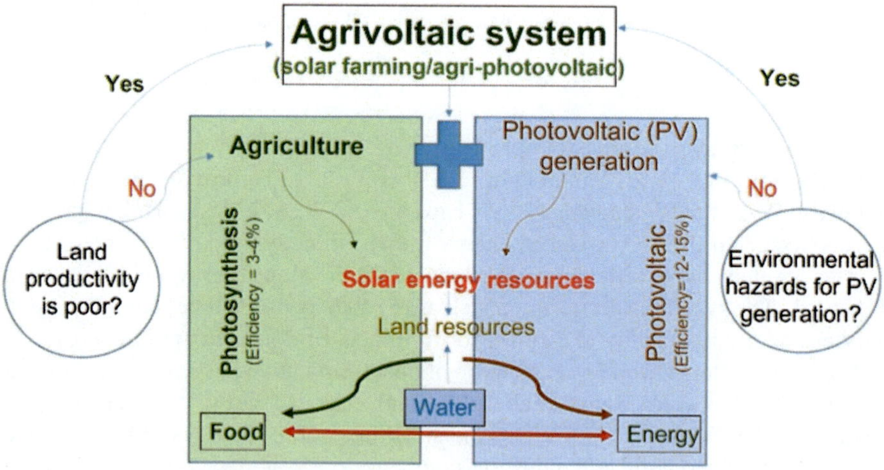

Fig. 5.7 Concept of agrivoltaic system

The AVS is most suitable in dryland regions of the world (Santra et al. 2018a; Barron-Gafford et al. 2019).

In addition to PV generation and food production, rainwater harvesting from top of the PV module of ground mounted system is also feasible (Santra et al. 2018a).

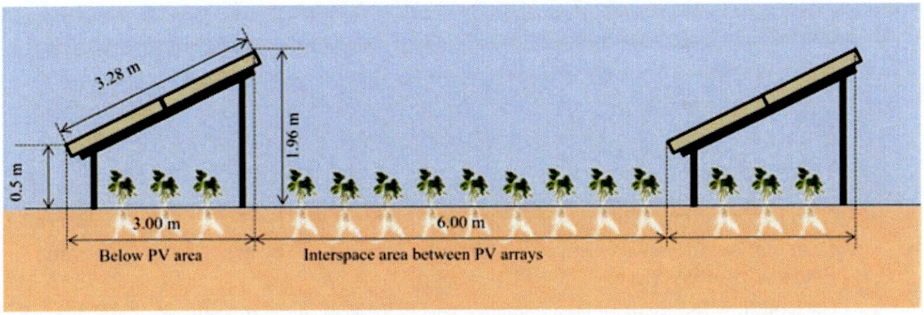

Fig. 5.8 Design of agrivoltaic system

Fig. 5.9 Agrivoltaic system installed at ICAR-Central Arid Zone Research Institute, Jodhpur

The harvested water is recycled for cleaning of PV module and for providing supplemental irrigation to crops in AVS.

Typically, the interspace area in between two arrays of the AVS is utilized for agricultural activity. Depending upon the height of mounting structure, the area below the PV module can also be utilized for agriculture purpose. A design of AVS model with two-row PV array is presented in Fig. 5.8. The details of the AVS design may be seen from Santra et al. (2021). The inclination of the PV module of the ground mounted system depends on the latitude of the place where it is installed. Generally, the inclination angle is kept as equal to the latitude of the location.

Following the design, field view of an AVS model installed at ICAR-central Arid Zone Research Institute is presented in Fig. 5.9. The established AVS model has the bottom row with full PV density, whereas top row has a PV density of 60% (Santra et al. 2021). As per the design, 400 kW_p AVS can be installed in 1 ha area.

Suitable crops to be grown in the interspace area of AVS depend on the climatic condition, the height of the mounting structure, availability of water and degree of tolerance of the crop to shade (Santra et al. 2020). If the height of mounting structure is low (e.g. 50 cm from ground level as in CAZRI AVS model), crops with low height (preferably shorter than 50 cm) need to be selected otherwise it will create

Table 5.4 Economics of agrivoltaic system

Sr. No.	Item	Value
1.	Area	1 ha
2.	Capacity (double row model)	400 kW_p
3.	Life cycle	25 years
4.	Cash outflow for ground mounted PV system (\$556 kW_p^{-1})	\$222,576
5.	Cash outflow for replacement cost of inverter (once in life cycle) (\$80 kW^{-1} of inverter)	\$39,746
6.	Cash outflow for repair and maintenance cost @0.1% of initial investment	\$223
7.	Cash outflow for crop cultivation (moong bean during kharif and isabgol during rabi) (\$ ha^{-1} y^{-1})	\$695
8.	Annual generation (@4 kWh/day/kWp with 1% decrease per year)	5,84,000 kWh
9.	Electricity sale price (Rs/kWh)	\$0.07
10.	Cash inflow from PV component (\$ ha^{-1} y^{-1})	\$38,686
11.	Cash inflow from crop component (moong bean and isabgol) (\$ ha^{-1} y^{-1})	\$1065
12.	Simple payback period	5.87 years
13.	Discounted payback period	10.40 years
14.	Internal rate of return	16%
15.	Net present value at a discount rate of 10%	\$92,940

(\$1 = INR 75.48)

shade on top of PV module and reduce PV generation. Availability of photosynthetic photon flux density (PPFD) or daily light integral (DLI) after interception by PV module at interspace and below PV area of AVS plays a critical role to select crops for AVS (Santra et al. 2021). Electricity generated by the PV modules of AVS may be sold to local grid through net metering system and the average PV generation rate at western Rajasthan is about 4–5 kWh per day per kWp installation.

The AVS is a potential solution to dryland farmers to increase the land productivity since rainfed-based crop production is risky because of uncertainty and scarcity of rainfall. It has been observed that land productivity may be improved by 41–62% by integrating PV generation and crop production in AVS (Santra et al. 2018b). Economic analysis of CAZRI AVS model carried out through LCC approach is presented in Table 5.4. Initial investment to install 400 kW_p capacity AVS in 1 ha is about \$222,576 considering the unit price of \$556 kW_p^{-1}. Annual income from selling of PV generated electricity is about \$38,686 per year with a decrease of 1% per year considering PV generation of 4 kWh per kW_p system per day and electricity tariff rate of 0.07 kWh^{-1}. Additional income of \$1060 per ha can be generated from crop yield (*Vigna radiata* during kharif and *Plantago ovata* during rabi) in the AVS. Life cycle cost analysis of the system shows simple payback period of 5.87 years, discounted payback period of 10.40 years and internal rate of return of 16% (Table 5.4).

In addition to the above economic analysis, a detailed technoeconomic analysis of AVS model with different designs and different cropping intensity is also available

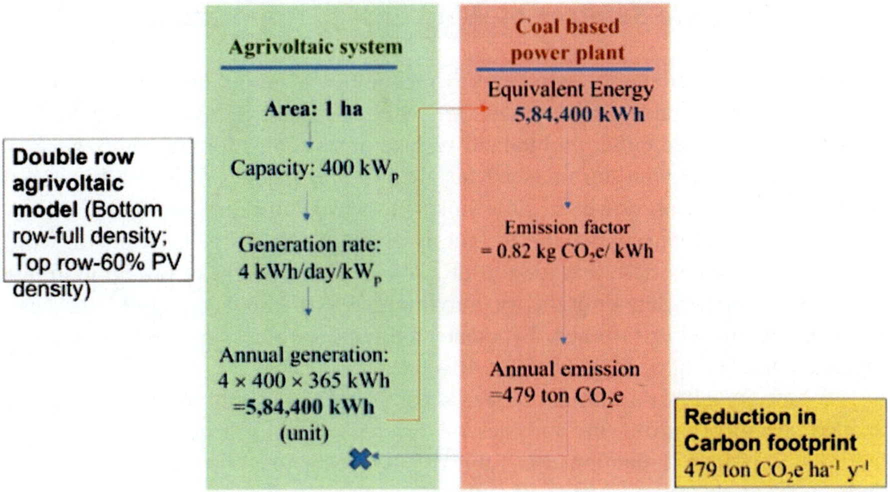

Fig. 5.10 Carbon footprint of agrivoltaic system

in Poonia et al. (2021). The net present value of the AVS model (refer Table 5.4) during its life cycle of 25 years is \$92,940, which can be considered as the provisioning services of AVS by providing electricity and crop yield.

The AVS model also provides several other ecosystem services. For example, adoption of AVS model in 1 ha can reduce the carbon footprint by 479 t ha^{-1} y^{-1} (Fig. 5.10). Thus, the regulating service of AVS is calculated as \$41,194 ha^{-1} y^{-1}, considering SCC value of \$86 t^{-1} CO2-eq (\$1 = INR 75.48).

Ground mounted PV system of AVS established in dryland is also expected to reduce soil erosion by wind action and thus provides supporting services, although data on reduction in soil erosion due to barrier effect of ground mounted PV structure is not available in literature. Moreover, the microclimate in AVS is modified by reducing the temperature and evaporation losses of water, which is expected to increase both PV generation and crop water productivity. All these benefits can be accounted as the supporting service of AVS model. Another important support service of AVS model is the rainwater cycling for cleaning the dust load of PV module. It is observed from field experiment that deposition of dust load on top of PV module reduces the PV generation and thus regular cleaning is essential at least once in a week. Deep groundwater sources in dryland are generally used for cleaning purpose. The harvested rainwater from AVS model can provide 40 ha-mm equivalent amount of water, which can be recycled throughout the year and thus can save \$661 y^{-1}, considering the cost for withdrawing 61 m deep groundwater @ \$0.32 ha-mm^{-1} (Meena et al. 2021).

5.3 Ecosystem Services of Wind Energy Applications

As compared to solar energy, wind energy technology is less utilized in agriculture. Generally, large capacity horizontal axis wind turbines (HAWT) (800 kW to 2.0 MW) are installed either on the land surface or seashores for electricity generation. Sometimes the wind turbines are installed in vast patches of agricultural land specifically in Gujarat, however, little link with agricultural production system. A field view of horizontal axis wind turbine is given in Fig. 5.11. Three-blade wind turbines are mostly installed at the top of a tower with a height of 60–80 m above ground level. As per Betz limit theory, maximum 59% of kinetic energy of wind can be harvested by wind turbines. Efficiency of wind turbines installed at western Rajasthan at a height of 80 m is about 40–44%.

The capacity of wind turbine varies as per the blade length and hub height. A detailed specification of wind turbines of different capacity installed in Jaisalmer district of Rajasthan is mentioned in Table 5.5. It is to be noted from the table that the blade length varies from 22 m to 55.9 m.

Requirement of land for installation of wind turbine depends on the blade length and the capacity of turbine. Generally, the wind turbines are installed as per the site suitability and therefore regular grid is not followed. A separation distance between

Fig. 5.11 Horizontal axis wind turbine installed at Jaisalmer, Rajasthan

Table 5.5 Technical specification of horizontal axis wind turbine

Model	Capacity	Blade length	Hub height	Total height	Swept area
Enercon E-53	800 kW	26.45 m	50–73 m	76.45–99.45 m	2198 m^2
Enercon E-48	800 kW	24 m	50–76 m	74–100 m	1810 m^2
Enercon E-44	900 kW	22 m	45–55 m	67–77 m	1521 m^2
Enercon E-82	2.00 MW	41 m	78–138 m	119–179 m	5281 m^2
Enercon E-92	2.35 MW	46 m	78–138 m	124–184 m	6648 m^2
Suzlon950	950 kW	32 m	73 m	105 m	3217 m^2
SuzlonS88	2.1 MW	44 m	80 m	124 m	6082 m^2
SuzlonS97	2.1 MW	48.5 m	90 m	138.5 m	7386 m^2
SuzlonS111	2.1 MW	55.9 m	90 m	145.9 m	9817 m^2
GamesaG80	2.0 MW	40 m	60–100 m	100–140 m	5027 m^2

two wind towers is maintained which is equal to 4–5 times of the blade length. For example, blade length of 2 MW turbine is approximately 50 m and thus a separation distance between two wind turbine towers is kept as 200–250 m. Thus, approximately 4–6.25 ha land area may be required to install a 2 MW turbine, which indicates a land requirement of about 2–3 ha MW^{-1}. Similarly, for installation of 800 kW turbine, land area of about 1–1.5 ha is required, which is almost equivalent to 2–3 ha MW^{-1}. Electricity generation from wind turbines is found highest during summer months, whereas solar PV system generates maximum during winter months. Thus, wind turbines and solar PV systems complement each other in electricity generation. The annual capacity utilization factor (CUF) of wind turbine varies from 17–19%, which indicates that it generates about 19% of the maximum amount of energy it could generate throughout the year. Considering the CUF value of 19% and land requirement of 3 ha MW^{-1}, the energy generation rate is 554.8 MWh ha^{-1}. Considering the wind energy tariff rate of Rs 2.40 kWh^{-1}, the provisioning service of wind energy generation is \$17,641 ha^{-1} y^{-1}. Apart from it, it provides regulating service of \$35,445 ha^{-1} y^{-1} by reducing emission of 455 t CO_2-eq ha^{-1} y^{-1}, which is otherwise released by coal-based thermal power plant to generate the same amount of energy. Establishment of wind turbines also provides other supporting services, e.g. providing job opportunity to local dwellers for operation and maintenance of turbine, establishment of social institutions (e.g. school, temples, community hall, etc.) through social welfare fund of the turbine owner, construction of road networks for transportation, etc.

Instead of harnessing wind energy at about 50–80 m above ground level (agl) using HAWT, surface wind energy can also be harnessed through installation of low-capacity vertical axis wind turbine (VAWT). One big advantage of the VAWT is the reduction of wind speed at leeward side of the turbine and thus reduces soil erosion. Although such turbines are not available in market but have potential in future considering dual purpose utility. A field view of 35 W capacity VAWT is presented in Fig. 5.12. The VAWT can generate electric energy at a minimum wind speed of about 1.5–2 m s^{-1} and the aerodynamic efficiency is about 20–30%. Considering the available wind speed >10 km h^{-1} for about 10 hours in a day

Fig. 5.12 A field view of vertical axis wind turbine

during summer months, 6–10 kWh energy may be generated in per ha per day using 20 VAWT units, which can be accommodated in 1 ha.

5.4 Ecosystem Services of Biomass-Based Energy Generation

5.4.1 Biogas Generation and Ecosystem Services

Biomass-based energy is an important energy source since it is renewable and widely available. It also has the potential to provide significant employment in the rural areas. In India, biomass materials used for power generation include bagasse, rice husk, straw, cotton stalk, coconut shells, soya husk, de-oiled cakes, coffee waste, jute wastes, groundnut shells, saw dust, etc. Biomass or agro-waste can be utilized to generate renewable energy, through either of these three following pathways: (i) gaseous fuels like biogas (methane), (ii) liquid fuels such as ethanol or pyrolysis oil and (iii) electricity. The thermochemical processes for conversion of biomass to useful products involve combustion, gasification or pyrolysis. The most used route is combustion. The advantage of combustion is that the technology is similar to that of a thermal plant based on coal, except for the boiler. The cycle used is the conventional ranking cycle with biomass being burnt in high pressure boiler to generate steam and operating a turbine with generated steam. The net power cycle efficiencies that can be achieved are about 23–25%. The exhaust of the steam turbine can either be fully condensed to produce power or used partly or fully for another useful heating activity. The latter mode is called cogeneration.

About 112 million tons of crop residues are surplus or available in India annually after accounting for its various traditional utilities. Anaerobic digestion of these crop residues can generate about 67 million tons of enriched and well-decomposed bio-manure, 12.88 million tons of bio-CNG worth $6.82 billion per annum @ $0.53 per kg. Anaerobic digestion of crop residues is also expected to moderate pollution, climate change and associated risks in agriculture. Employment of 0.5 million in the biogas plants at primary level and 0.7 million at the secondary level have been estimated required for baling, collecting and handling the biomass straw. Bio-CNG generation and its benefits can be further multiplied by cogeneration. An integrated system of anaerobic digestion of crop residues, other biomass, animal excreta, sewage, municipal solid wastes, industrial, processing and food wastes can be designed to optimize the bio-gas and manure production throughout the year. However, it has to be ensured that the bio-manure is recycled to agricultural fields.

5.5 Future Road Map

Despite several advantages of renewable energy use in agriculture, as evident from the several ecosystem services, adoption of these technologies in farmers' field and rural hinterland is far from satisfactory. Therefore, there is need for suitable policy interventions to promote adopted farmers through incentives calculated based on ecosystem services of renewable energy systems and devices. The calculation of prices of ecosystem services as presented above is based on data and parameters from isolated field experiments. Even, the ecosystem services of few renewable energy devices could not be calculated because of lack in quantified data and parameters. Moreover, there is huge scope to include several other renewable energy-based technologies in agriculture having high ecosystem services. Based on these observations, the following future actions may be noted:

(i) Data and parameters required for calculation of price of the ecosystem services need to be generated from multiple field experiments at several locations across agroclimatic zones.
(ii) Ecosystem services of each renewable energy device need to be considered for providing monetary incentives to adopted farmers. This will not only create an eco-friendly and climate resilient production system but also will increase farmers' income.
(iii) Farm mechanization using renewable sources, e.g. solar powered tractors and vehicles, is another futuristic avenues, which has huge role in increasing the share of renewables in agriculture.
(iv) PV greenhouse is another option to generate and utilize the power required to maintain the optimum enclosure environment for plant growth and to provide

irrigation to crops. Excess energy may also be used for other mechanized operations for sorting and grading of agricultural produces within the system.

(v) Solar-wind hybrid system may be adopted at field boundaries of the agricultural field. This will not only generate electricity but will also protect the top nutrient rich soil from wind erosion.

(vi) Efforts are required to quantify cultural services of renewable energy use by farmers, e.g. off-time usage of solar PV system for meeting the domestic need and recreational activities (e.g. light, television, fan, radio, battery charging of smartphones)

(vii) On-farm post-harvest processing and value addition of agricultural produces has potential to reduce transportation loss of raw produces to markets and to fetch better market price after value addition. All these post-harvest processing and value addition can be operated through renewables.

5.6 Messages

In this chapter, ecosystem services of several renewable energy devices are calculated in terms of price. The required data and parameters to calculate four major ecosystem services, e.g. provisioning service, supporting services, regulating services and cultural services, often are not available. Several assumptions have been made based on the field experience and observations. All these parameters may vary as per locations and situations. Therefore, it is suggested to collect data and parameters as per the requirements of local situations to calculate ecosystem services using the equations and methodology as mentioned in this chapter. Adoption of renewable energy devices sometimes may reduce the provisioning services, which is a tangible outcome and hence is highly rated while considering for adoption of that renewable energy-based technology. Under such situation, trade-offs between positive ecosystem services (e.g. increase in regulating services, support services and cultural services) and negative provisioning services need to be calculated. Prices of the ecosystem services of wind energy and biomass energy-based systems could not be fully calculated in this chapter here because of lack of data and their limited use in agriculture. Doubling farmers' income can be achieved if the farmers are given ecosystem service based monetary incentives in case, they adopt renewable energy system.

References

Barron-Gafford GA, Pavao-Zuckerman MA, Minor RL, Sutter LF, Barnett-Moreno I, Blackett DT, Macknick JE (2019) Agrivoltaics provide mutual benefits across the food–energy–water nexus in drylands. Nat Sustain 2(9):848–855. https://doi.org/10.1038/s41893-019-0364-5

Central Electricity Authority (2018) CO_2 baseline database for the Indian power sector, user guide, version 13. Ministry of power, Govt of India. http://www.cea.nic.in/reports/others/thermal/tpece/cdm_co2/user_guide_ver13.pdf

Elsoragaby S, Yahya A, Mahadi MR, Nawi NM, Mairghany M (2019) Energy utilization in major crop cultivation. Energy 173:1285–1303

IEA (2021) Global energy review 2021. IEA, Paris. https://www.iea.org/reports/global-energy-review-2021

IRENA (2021) Renewable capacity statistics 2021 international renewable energy agency (IRENA), Abu Dhabi

Kumar M, Reddy KS, Adake RV, Rao CVKN (2015) Solar powered micro-irrigation system for small holders of dryland agriculture in India. Agric Water Manag 158:112–119

Meena HM, Santra P, Singh RK (2021) Quantification of water productivity and economics of irrigation in summer clusterbean (*Cyamopsis tetragonoloba*) in hot arid region of India. Ind J Soil Conser 49(2):106–111

NAAS (2018) Renewable energy: a new paradigm for growth in agriculture. Strategy Paper No. 10, National Academy of Agriculture Sciences, New Delhi, 20p. http://naas.org.in/spapers/Strategy%20Paper%20No.%2010.pdf

Nahar NM (1994) Design, development and testing of a large-size solar cooker for animal feed. Appl Energy 48:295–304

Nahar NM (1998) Design, development and testing of a novel non-tracking solar cooker. Int J Energy Res 22(13):1191–1198

Nahar NM, Gupta JP, Sharma P (1996) A novel solar cooker for animal feed. Energy Convers Manag 37(1):77–80

Nahar NM (2001) Design, development and testing of a double reflector hot box solar cooker with a transparent insulation material. Renew Energy 23(2):167–179. https://doi.org/10.1016/S0960-1481(00)00178-6

Nugent D, Sovacool BK (2014) Assessing the lifecycle greenhouse gas emissions from solar PV and wind energy: a critical meta-survey. Energy Policy 65:229–244. https://doi.org/10.1016/j.enpol.2013.10.048

Pande PC, Nahar NM, Chaurasia PBL, Mishra D, Tiwari JC, Kushwaha HL (2009) Renewable energy spectrum in arid region. In: Kar A, Garg BK, Singh MP, Kathju S (eds) Trends in arid zone research in India. CAZRI, Jodhpur, pp 210–237

Pande PC, Santra P, Singh AK (2016) Potential of utilizing solar energy for reducing carbon emission. In: Bhatt RK, Burman U, Painuli DK, Singh DV, Sharma R, Tanwar SPS (eds) Climate change and agriculture: adaptation and mitigation. Satish Serial Publishing House, Delhi, New Delhi, pp 401–416

Pande PC, Singh AK, Ansari S, Vyas SK, Dave BK (2003) Design development and testing of a solar PV pump based drip system for orchards. Renew Energy 28(3):385–396

Pande PC, Thanvi KP (1988) Design and development of a solar cooker cum drier. Int J Energy Res 12(3):539–545

Pande PC, Thanvi KP, Nahar NM, Ramana Rao BV (1980) Multipurpose solar energy device. Ann Arid Zone 19(4):525–528

Poonia S, Jat NK, Santra P, Singh AK, Jain D, Meena HM (2021) Techno-economic evaluation of different Agri-voltaic designs for the hot arid ecosystem of India. Renew Energy 184:149–163. https://doi.org/10.1016/j.renene.2021.11.074

Poonia S, Singh AK, Santra P, Jain D (2017a) Performance evaluation and cost economics of a low cost solar dryer for ber (*Zizyphus mauritiana*) fruit. Agric Eng Today 41(1):25–30

Poonia S, Singh AK, Santra P, Nahar NM, Mishra D (2017b) Thermal performance evaluation and testing of improved animal feed solar cooker. J Agric Eng 54(1):33–43

Rao CS, Srinivas I, Adake RV, Santra P, Reddy BS, Kumar M, Rao KV, Reddy KS, Yadav OP, Saxena MC (2017) In: El-Beltagy A, Saxena MC (eds) Utilization of renewable energy sources in dryland system, in: twelfth international dryland development conference on sustainable development of drylands in the post 2015 world. International dryland development commission (IDDC), Egypt, pp 537–552

Ricke K, Drouet L, Calderia K, Tavoni M (2018) Country-level social cost of carbon. Nat Clim Chang 8:895–900. https://doi.org/10.1038/s41558-018-0282-y

Santra P (2015) Scope of solar energy in cold arid region of India at Leh Ladakh. Ann Arid Zone 54(3&4):109–117

Santra P (2021) Performance evaluation of solar PV pumping system for providing irrigation through micro-irrigation techniques using surface water resources in hot arid region of India. Agric Water Manag 245:106554. https://doi.org/10.1016/j.agwat.2020.106554

Santra P, Meena HM, Yadav OP (2021) Spatial and temporal variation of photosynthetic photon flux density within agrivoltaic system in hot arid region of India. Biosyst Eng 209:74–93. https://doi.org/10.1016/j.biosystemseng.2021.06.017

Santra P, Pande PC, Singh AK (2017a) Solar PV pumping system for irrigation in agriculture. In: Ghoshal MK (ed) Renewable energy technologies. Narosa Publishing House, New Delhi, pp 277–292

Santra P, Pande PC, Kumar S, Mishra D, Singh RK (2017b) Agri-voltaics or solar farming: the concept of integrating solar PV based electricity generation and crop production in a single land use system. International Journal of Renewable Energy Research 7(2):694–699

Santra P, Pande PC, Singh AK, Kumar P (2016) Solar PV pumping system for irrigation purpose and its economic comparison with grid-connected electricity and diesel operated pumps. Indian Journal of Economics and Development 4(4):1–7

Santra P, Singh RK, Jain D, Yadav OP (2018b) Agri-voltaic system to enhance land productivity and income. Indian Farming 68(9):108–111

Santra P, Singh RK, Meena HM, Kumawat RN, Mishra D, Jain D, Yadav OP (2018a) Agri-voltaic system: crop production and photovoltaic-based electricity generation from a single land unit. Indian Farming 68(01):20–23

Santra P, Singh RK, Meena HM, Kumawat RN, Mishra D, Machiwal D, Dayal D, Jain D, Yadav OP (2020) Agri-voltaic system for crop production and electricity generation from a single land unit. In: Singh S, Ramadesigan V (eds) Advances in energy research, Vol 1. Springer Proceedings in Energy. Springer, Singapore, pp 45–56

Santra P, Singh RK, Poonia S, Jain D (2019) Solar energy in agriculture: principles and applications. New India Publishing Agency, New Delhi, p 274

Singh AK, Poonia S, Santra P, Jain D (2020) Ensuring energy and food security through solar energy utilization. In: Singh P, Singh R, Srivastava V (eds) Contemporary environmental issues and challenges in era of climate change. Springer, Singapore, pp 199–218

Singh H, Mishra D, Nahar NM (2002) Energy use pattern in production agriculture of a typical village in arid zone, India–part I. Energy Convers Manag 43(16):2275–2286

Singh H, Mishra D, Nahar NM (2004) Energy use pattern in production agriculture of a typical village in arid zone–part III. Energy Convers Manag 45(15–16):2453–2472

Singh H, Mishra D, Nahar NM, Ranjan M (2003) Energy use pattern in production agriculture of a typical village in arid zone India: part II. Energy Convers Manag 44(7):1053–1067

Swami V, Chauhan DK, Santra P, Kothari K (2016) Design and development of solar PV based power sprayer for agricultural use. Ann Arid Zone 55(1&2):51–57

Crop Insurance Based on Payment of Ecosystem Services

6

6.1 About Crop Insurance

Agriculture co-exists with risks from erratic weather; variations in yields, prices; changes in government policies, international markets, labour availability, and additional issues that influence agricultural activities and set off ample fluctuations in farm earnings many of which are beyond the control of farmers (Kiran and Umesh 2015). Under the situations of risks and uncertainties, various schemes have been devised and operated over time to protect the farmers from losses that included guaranteed prices, subsidized inputs, credit and crop insurances. Crop insurance is a kind of compensation policy that protects farmers against unexpected loss of anticipated crop yields or returns from produce sold at market, i.e. crop insurance is related with either crop yield or crop revenue. Crop yield insurance compensates the insured farmers for yield losses, while the revenue insurance compensates for losses of revenue below a threshold level in case of unforeseen events. Both types of crop insurance are means to aid in risk management for producers due to unexpected events. Crop insurance provides self-sufficiency and confidence among farming community, while in the event of crop loss they can claim for compensation from authorities. It protects the farmers from the jolt results from crop loss by reassuring farmers' safeguard against natural perils, and to deal with the changes in climate (Aryal et al. 2020). During recent years in India, both the state and central governments introduced so many crops insurance schemes as a safeguard measure to protect farmers against yield loss or adverse weather events.

Information are available for farmer's level risk characteristics, which influence selection of the scheme, evidence is scarce on valuations of individual attributes of crop insurance. In the Indian context, there are top-down approach for risk assessment, which hardly makes the local stakeholders participate in designing the product and led to lower recognition among target groups (Feder et al. 1985). The question arises about low voluntary participation in spite of very low premiums that are fairly reasonable. Whether quality of insurance product is restraining rather than farmers'

P. Bhattacharyya et al., *Pricing of Ecosystem Services in Agriculture: A Basis of Crop Insurance*, https://doi.org/10.1007/978-981-19-4416-1_6

capability to pay and the attributes of an insurance scheme, which are attractive to the farmers, with respect to their specific requirements and many more questions are there. Nothing of these can be replied reasonably without valuation of farmers' willingness-to-pay (WTP) for inclusive crop insurance scheme. Meanwhile, any effort for designing a large-scale insurance policy intended for more acceptance remains not only unreachable, but also absence of the required market intelligence for maximizing their rate of success. This particular gap can be filled through indication on farmers' choices for crop insurance policy features, based on some discrete choice experiment (DCE) done across regions to measure their valuation in totality for a crop insurance product. Knowledge of attributes of an insurance product that are more significant to farmers aid the policymakers for making the needed modifications in existing scheme and increase its overall efficacy and usefulness.

For compensating any expenses made for production or conservation activities, the concepts of Payments for Ecosystem Services (PES) are evolved with the arrangements in which the recipients of environmental goods/services will pay for watershed protection and conservation of forest resources for sequestration of carbon, beautification of the landscape, etc. for rewarding the farmers and environmental stewards with payments or subsidies. Making arrangements for payments in exchange of the benefits received from forests, fertile soils and other natural resources is a way to know their worth and warrant that the benefits would continue in future. The PES is an encouraging conservation strategy, which can benefit sellers, buyers and expand the resource base, which can be linked to future insurance strategy.

6.2 Evolution of Crop Insurance Schemes in India

An expert committee constituted by Government of India under Prof. Dharam Narain as Chairman to assess the likelihood of starting of crop insurance schemes in India and the committee endorsed for initiation of insurance programme for farmers. Based on the endorsements, General Insurance Corporation (GIC) was set up by Government of India during 1973. The GIC started a crop insurance programme in Gujarat during the same year on pilot basis, which was then protracted to Andhra Pradesh, Tamil Nadu and West Bengal. During the year 1985, the first ever complete crop insurance programme, namely, Comprehensive Crop Insurance Scheme (CCIS), was introduced by GIC in different states. The scheme directed to offer protection to the farmers from risk, who availed loans and the sum insured was limited to USD $808.41 (INR. 10,000) per farmer regardless of the loan amount. The premium was decided at 2% (of sum assured) for cereals and 1% for oilseeds and pulses. In spite of these efforts, the scheme could not realize its envisioned objectives because of the faults like narrow crop coverage, limit on sum insured, and it was available for loanee farmers only. The Government started the National Agricultural Insurance Scheme (NAIS) during the year 1999 with wide crop coverage, making available for both loanee and non-loanee farmers, with the inclusion of loanee farmers mandatorily. For monitoring the scheme effectively, Agriculture Insurance Company was established in the year 2002.

A weather index-based Crop Insurance Scheme was piloted across India from *kharif* 2007, to judge the efficiency of weather-based crop insurance as a method to tackle the restrictions of NAIS like adverse selection, moral hazards, etc. Weather-related factors are used as an index for compensation of crop loss using simulation models of crop yield and related weather parameters. Farmers get compensated when weather parameters change above or below the normal value. The scheme was destined with many restrictions like availability of automatic weather stations and parametric weather information.

Later, NAIS was altered and re-launched as Modified National Agricultural Insurance Scheme (MNAIS) and implemented again in few states, on pilot basis. Unlike NAIS, MNAIS works in the regime of actuarial premium and has a broad definition of risk, which comprises cyclones and area not sown (due to hazards). In 2013, NAIS, WBCIS and Coconut Palm Insurance Schemes (CPIS of 2009) were amalgamated into an inclusive scheme, i.e. National Crop Insurance Programme (NCIP). This scheme was in action in all the districts since *kharif*, 2014.

There are ample evidence on the nature and impact of agricultural insurance, but there is a gap in information regarding farmer's preferences in comparison to other methods of managing risks (Skees 2000; Jensen et al. 2018). In the situation of high acquaintance to manifold risks, particularly in developing nations, large-scale, subsidized multi-peril indemnity-based crop insurance programmes have become integral to governmental strategy for mitigation of agricultural risk (Hazell and Varangis 2020). However, take-up rates of these programmes remained meagre in absence of support through government subsidies (Yu and Sumner 2018; Feng et al. 2020). Even in advanced nations also, there is not ample indication to advocate that adoption of risk mitigation strategies among farmers is large enough to pay for entirely private actuarial premiums (Goodwin 2001; Smith and Glauber 2012). Earlier indemnity-based insurance programmes were having many constraints, including asymmetry in information about adverse selection and moral hazard (Hazell 1992; Morduch 2006; Barnett and Mahul 2007; Miranda and Farrin 2012; Wu et al. 2019) and aversion of ambiguity (Elabed and Carter 2015). They were also susceptible to added problems like high managerial expenses (for example, expenditure to assess the loss), and covariate risks, which either converted to higher risks of insolvency or upsurge of re-insurance cost. All these problems are more noticeable in emerging nations where information asymmetry, knowledge gaps and other organizational and functioning issues are even higher. Furthermore, even with substantial exploration, scanty information is available to advocate that traditional crop insurance disturbs welfare of farmers (Hazell 1992; Smith and Watts 2010; Fadhliani et al. 2019). Existence of numerous glitches in crop insurance markets can cause irregularities in farmers' choices during coverage periods (Babcock 2015; Huo et al. 2018). Though insurance against yield damages from multiple perils is abundant since long periods, acceptance rates have remained basically unsatisfactory. The most recent experience comes from the *Pradhan Mantri Fasal Bima Yojana* (PMFBY), launched during 2016 replacing the existing National Agriculture

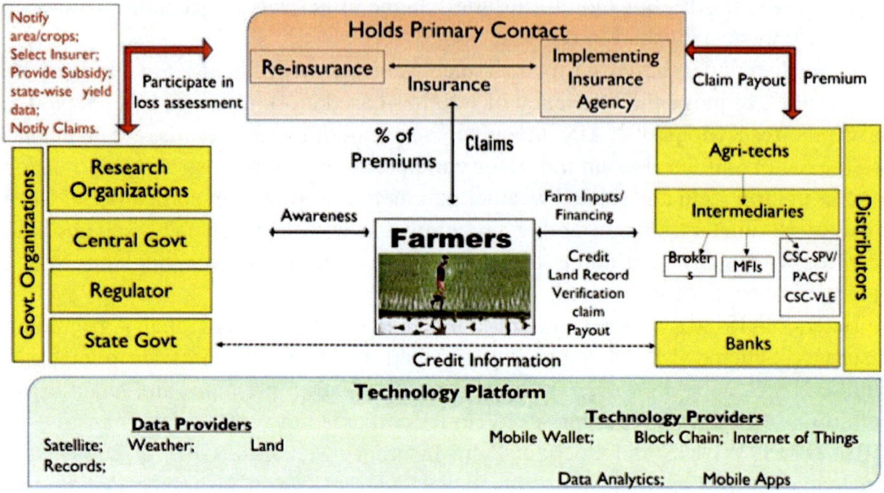

Fig. 6.1 Crop insurance ecosystem

Insurance Scheme (NAIS). From the time when it launched, about 55 million farmers have been insured, enrolling more than the preceding scheme by almost 40%, making it one of the biggest crop insurance programme in the world. But, regardless of the low premium and attractive terms, limited number of farmers willingly bought this insurance product under the new policy framework. Merely 25% of covered farmers obtained it by their choice; the rest 75% were covered by virtue of receiving loan. The crop insurance ecosystem in India, which is a diverse system of participants that together serve the farmers is illustrated in Fig. 6.1.

6.3 Weather-Based Crop Insurance Scheme (WBCIS)

6.3.1 Features and Design of Weather Index

Weather index is a quantitative index, which correlates the shortfall in yield of crops with weather parameters (Clarke et al. 2012). Generally, the weather parameters which can be measured easily are considered for forming the index like temperature, humidity, rainfall, etc. Few of the significant attributes of such index are: (i) easy to measure, (ii) unbiased, (iii) transparent, (iv) can be easily verifiable, (v) timely noted and reported, (vi) consistent, (vii) experienced over a large geographical area and (viii) extremely interrelated with risks. The WBCIS works on area approach (in contrary to individual approach for earlier schemes) and for payment, a 'Reference Unit Area' (RUA) is chosen as insurance unit and is associated with a Reference Weather Station (RWS) for getting weather data for index calculations. The RUA is informed about 2 months prior to the starting of crop season and the insured farmers belonging to RUA are considered equally for any payout due to irregularities of weather. 'Triggers' or threshold values are fixed based on temporal data on weather parameters, which are used for compensation of likely loss in crop yield. If the weather events of the RUA within a specified period of time deviate from the trigger value, farmers received the compensation. For instance, during hot waves, for every degree Celsius increase in temperature from threshold, a definite sum is paid as compensation antedating shortfall in productivity due to irregularities.

6.3.2 Crops Covered and Premium Under WBCIS

Cereals, pulses, millets, oilseeds and annual commercial/horticultural crops cultivated during the *rabi* season are protected under the WBCIS. The major *rabi* crops are wheat, barley, potato, onion, gram, mustard, lentil, coriander, cumin, fenugreek, isabgol, etc., whereas the *kharif* crops covered under this scheme include rice, bajra, maize, jowar, ragi, groundnut, soybean, pulses, cotton, etc. The State Level Coordination Committee on Crop Insurance (SLCCCI) notifies the crops covered in each reference unit area. Premium rates are decided on the anticipated deficit in yield using past weather data of RUA. But the rate of compensation is fixed, irrespective of quantum of losses for the individual farmers. The rate of premium for the farmers are as in Table 6.1.

There are some advantages of weather-based crop insurance, as:

- *Minimum requirement of data:* Other insurance products necessitate lots of data for calculation of average yield and assessment of risks, where index insurance removes requirement of individual farmer's data, instead use homogenous area approach.
- *More transparent:* Weather index-based insurance is more transparent in payout process; all stakeholders have access to the index as well as payment-related information. The weather-based data is simple and no disputes arise between the farmers and the agency in terms of compensation, which creates belief and trust among them, which make certain for more involvement of farmers.
- *Loss assessment is easy:* After any unfavourable weather incident is happened, field loss assessment becomes the major task in previous insurance schemes, which depend on field verification physically for loss assessment. In this scheme, such calculation is not essential as index values are adequate for chalking out payout plans, thus eliminating the visit to field for assessment of loss.
- *Lower operational expenses and quick compensation:* Man-power requirement is less in weather index insurance, hence operational cost got reduced. Further, settlements done automatically for all the farmer as soon as the index values deviate from the thresholds that got measured using weather station data.

Table 6.1. Rate of premium for different crops

Sl. No.	Crops	Premium payable by the insured farmer
Rabi season		
1. Wheat		1.5% or actuarial rate, which-ever is lower
2. Other crops (other cereals, millets, pulses, oilseeds)		2.0% or actuarial rate, which-ever is lower
Kharif season		
1. Bajra and oilseeds		3.5% or actuarial rate, which-ever is lower
2. Other crops (cereals, pulses and other millets)		2.5% or actuarial rate, which-ever is lower
Annual commercial/horticulture crops		
Premium slab		Subsidy/premium
1.	Up to 2%	No subsidy
2.	>2–5%	25%, limited to minimum net premium of 2% payable by farmer
3.	>5–8%	40%, limited to minimum net premium of 3.75% payable by farmer
4.	>8%	50%, limited to minimum net premium of 4.8% and maximum 6% payable by farmer

However, there are some limitations also, like:

- *Basis risk*: Basis risk is the deviations of payment received by the farmer from the economic loss experienced. Sometimes the situation may arise where payment may be made without actual loss or there may be fewer or no compensation provided even actual losses have been happened.
- *Truthfulness of weather data*: If the weather station data tampered by any means, then for the resultant index may be biased. Sometimes, agency itself indulges this practice for their own benefit.
- *Limited perils*: Weather index insurance contemplates single or maximum two weather perils, parting other hazards that also causes economic loss like fluctuations in price level, pests and diseases attack, etc.
- *Shortage of weather data and weather stations*: Temporal as well as current period weather data are obligatory for effective operation of weather index insurance. There is lack of historical data and requisite number of weather stations for maintaining a sensibly good relationship between the index and the homogenous area.
- *Technical expertise*: Designing of index-based insurance products using the weather information and preserving the weather stations for lengthier period necessitate a definite skill.

- *Scalability*: Indices created for particular crop of a particular area, which falls in a reference unit area of the weather station that cannot be generalized for other area.

6.4 'Prime Minister Crop Insurance Scheme, India (*Pradhan Mantri Fasal Bima Yojana, India*): One Nation—One Scheme theme'

PMFBY started since 13th January, 2016 by the central government substituting former two schemes, viz. NAIS and Modified NAIS by incorporating the finest attributes of preceding schemes and removing their shortcomings/weaknesses. This scheme which is administered by the Ministry of Agriculture and Farmers' Welfare executed in most of the states initially with the support of the particular state governments. The scheme aimed at offering an improved insurance product to the farmers with less premium and higher insurance coverage. The new scheme removed the limit on sum-insured and covered additional risks like post-harvest damages, area not sown and many localized catastrophes like cyclones, which were not in the list for earlier insurance schemes. Creation of awareness is given top priority under this scheme. The chief advantages of the scheme are: (i) coverage of all crops, (ii) uniform rate of premium for all farmers (2% of sum assured in *kharif* and 1.5% in *rabi*), (iii) no cap on claims of sum insured, (iv) payment of claims up to 25% of sum insured, if sowing is not done due to adverse weather, (v) due to localized calamities like inundation, hailstorm, landslide, etc. assessment is done at individual farm level, and (vi) application of improved technologies like remote sensing and drones to complement the actions for quick claim settlement.

Crop insurance provides assistance to farmers to cope with risk through compensation due to crop loss and the Government by sinking the load on account of disaster payments to farming sector. In spite of its significance, crop insurance in India has not expanded much among the farmers due to indifference owing from low awareness. Though there are many issues and constraints, the new scheme became a boon to the farmers (Fig. 6.2).

6.5 Technology-Based Crop Insurance in Agriculture

6.5.1 Need and Choices of Technology Adoption in Crop Insurance

The PMFBY is a transformative scheme aimed to offer insurance coverage and economic backing to farmers for catastrophe to any of the notified crop, uncultivated area and post-harvest losses owing from natural adversities, pests, diseases infestation, etc. for stabilizing the returns from agriculture, and to inspire the farmers for adopting newer farming methods. The scheme identifies the requirement for technological intercessions in crop insurance for making the insurance instrument further effective, transparent and farmer-centric. Seeing the intricacies related with Indian agriculture like small and fragmented landholdings, high eco-geographical

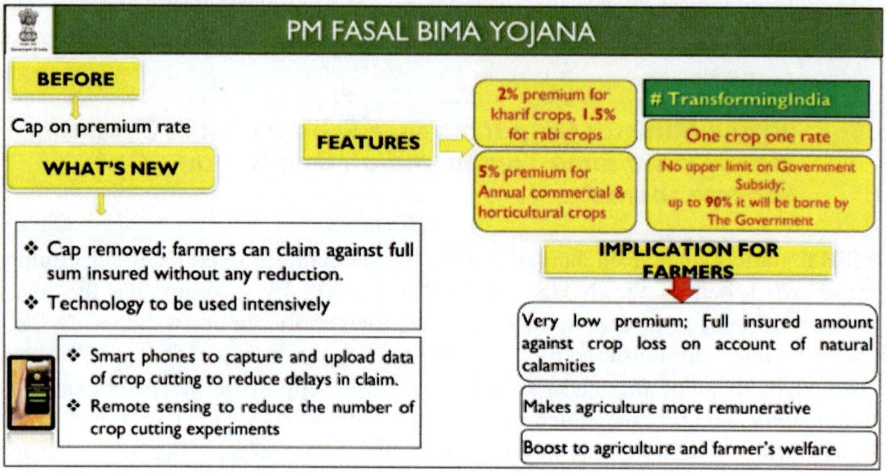

Fig. 6.2 Improved version of Prime Minister Crop Insurance Scheme, India

unevenness, yield variations and weather abnormalities, it is imperious that technologies can be efficiently adopted for increasing the usefulness and efficacy of the insurance sector.

For efficient operation of the PMFBY, some technological choices have been projected like smartphones, digital photography, remote sensing technologies (satellite and Unmanned Aerial Vehicles—UAVs), new statistical methods and modelling approaches, and IT/ICTs. Presently, IT/ICTs are in use for enrolment and other operational issues at modest level. However, there are only some discrete studies regarding use of technologies for estimation of yield, assessment of loss and product design but the evidence base of these technologies is limited to support their countrywide application. There is no comparative appraisal of different technologies available and the researchers as well as insurance players have dissimilar opinions regarding their use and efficacy. The main technology providers in the area of remote sensing are research and academic organizations and private companies, whereas IT/ICT and UAV technology suppliers are mostly private companies. Other technologies such as integrated assessment modelling, and statistics are still in the domain of national and international research organizations.

A Task Force on 'Use of Technology for Agriculture Insurance' was constituted by the NITI Aayog, who made discussion on this theme with almost 100 prominent experts. Increasing use of technology in agriculture insurance was advocated from research institutes, universities, insurance industry, the government, banks and other stakeholders for recommending the main technologies for active execution of the PMFBY. The task force documented numerous organizational and institutional constraints outside its mandate that required to be talked for efficient operation of the PMFBY besides use of technology. The task force instituted five sub-groups for doing a detailed assessment of technological choices and to get feedback from stakeholders on various issues. The sub-groups were on: (1) Remote sensing and drones; (2) Decision support systems, crop modelling and integrated approaches; (3) IT/ICT in Insurance; (4) Crop cutting experiments (CCEs) and (5) Technologies for livestock and aquaculture insurance. After considering the reports from all five thematic sub-groups, few solid recommendations on improving technology use in agricultural insurance were made by the task force as detailed below:

1. For increasing awareness and participation, the governments should immediately develop 24 × 7 links with the payment gateways/e-wallets through software/apps for facilitating and easing the registration process, online payment of premium and issue of e-receipts instantly. This can be linked with the National Crop Insurance Portal, which will enable to access data to the state government departments, bankers and insurance companies.
2. Campaigning should be done for enrolment during the crop season using ICT tools such as voice blasts, IVRS and SMS and participatory videos, etc. for awareness creation.
3. PMFBY guidelines should be hassle-free to reduce farmers' inconvenience with the enrolment process.
4. Digitization of geo-referenced documents of landholding is essential to avoid moral hazards problem. This work is currently being done in the states at a varied pace. Preparation of a strong geo-referenced and frequently updated cadastral map and its linking with land records need to be augmented and completed in a timely manner.
5. The task force indicated that though CCEs are crucial for the insurance scheme, it is not followed appropriately because of the scientific, financial, institutional and operational hurdles linked with them. Furthermore, not necessarily they would deliver correct appraisals of yield loss over insurance unit area. It is to be noted that an amalgamation of various options like digital photography, remote sensing, statistical methods, integrated crop modelling, etc. can deliver an impartial estimate of loss in crop yield with much less costs.
6. Though there are several satellites for supporting crop insurance sector, the task force endorsed for a committed constellation of few satellites of moderate to high resolution (10–30 m) with 10-days frequency and with multispectral optical sensors for increasing the accuracy of crop yield estimates/loss calculation at the village scale.
7. For localized climatic events like hailstorms, flash floods, landslides, and post-harvest losses, where CCE data do not play a role, required to be reinforced by mobile-based Apps.
8. Past yields are required for computing the likely risks and premium, for which data normally not obtainable at the preferred scale. Based on requirement and urgency, government in association with AICI should launch projects for

developing estimates of threshold yields of main crops at the wanted insurable unit scale for the past 10 years at least.

9. Suitable agro-ecological region-specific weather triggers should be developed for the area, where WBCIS is to be implemented, using integrated approach of historical weather data, statistics, crop models and remote sensing data, which can maximize the satisfaction of farmers.
10. Using seasonal and short-term weather forecasts, satellite images, temporal databases and current weather data, an independent tool can be prepared for providing a double trigger product for calculation of mid-season claim and 'on-account' payment to the insured farmers.
11. Considering the growing importance of livestock and aquaculture in Indian economy we should have efficient insurance policies for them. However, there exist some institutional and policy issues associated with insurance for livestock and fisheries sector that must be addressed concurrently.
12. The government should contemplate for launching a devoted process either from the existing institutions or by creating a new agency for coordinating between different stakeholders comprising farmers, industry, various government departments, banks and technology providers associated to insurance in agriculture and allied sectors. The same agency could also be mandated for collection, storage and transfer of data among the stakeholders.

6.5.2 Technology Enablers and Use Cases

The key technology enablers (Fig. 6.3) to complement the above issues and overcome the challenges mentioned in the previous sub-section are as below:

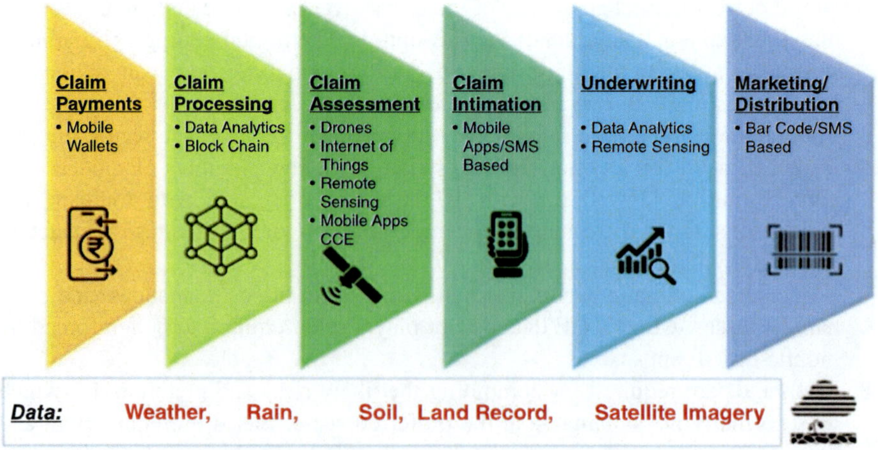

Fig. 6.3 Technology enabler in crop insurance

1. *Data and analytics*: The available data analytics use cases can play an alternating role through the crop insurance value chain, few scenarios are explained below:
   - *Effective CCEs*
   The following information can be adopted for digitization of the insurance process: (i) satellite imagery, (ii) vegetation indices, (iii) real-time data from weather and rain gauge stations, (iv) soil and crop data, (v) time series video of crop growth, (vi) census data, etc.
   - *Discrepancies in area sown and insured*
   The digital land records are geo-coded latitude/longitude village shape files, which can be used to ascertain the plots for CCEs, mapping of plot boundaries, and identification of mis-matches in area/crops sown, if any.
   - *Multiple insurance coverage for one plot*
   Based on the digital land records, case of excess or double insurance against the insurance application can be identified when integrated with distributor and insurer systems.
   - *Faster claims processing*
   Integration of insurer data with real-time weather and satellite data providers can assist insurers the claim intimation process to automatic and the claim amount can be directly transferred to farmers through digital banking.
   - *Underwriting*
   The satellite data can also be used for monitoring the crop health through current and relative analysis of vegetation status for assessing the risk profile of the region/plot.
2. *Drones*: Drones can be used for assessing the crop damage and thus speed up the claim settlement process. The temporal data collected can be used to progress the underwriting process. The images captured can also be overlaid on digital maps of states for identifying the farms and crops cultivated.
3. *Block-chain*: Participants in the crop insurance ecosystem can be considered as nodes on the block-chain. The contract established between the insurer and farmer and the data providers involved as trusted third parties. In case of an eventuality, actions will be activated and claims settled instantly on the block-chain.
4. *Mobile applications claim intimation:* This contains mobile applications using for intimation of losses and includes location-specific details as well as pictures of crop damage.
5. *Conducting digital CCEs:* Here, mobile applications can cover the locations, for which photographs are available with respect to CCEs.
6. *Internet of Things (IoT)*: Using the sensors for monitoring weather conditions, soil health, and crop health can produce data points, which can be used for performing analytics for real-time control.
7. *Distribution using digital channels*: Digital channels allow agro-dealers for providing packetized insurance cover. Post-buying, the farmer can use bar-coded scratch cards or SMS to avail insurance cover.

Above-mentioned technology enablers can accomplish a revolutionary role in the crop insurance value chain, which can provide accurate and quicker results.

6.5.3 Institutional Mechanisms for Technology Use in Crop Insurance

Crop insurance is an intricate subject due to largeness and dispersed population of marginal and small landholders who do farming in diverse agro-ecologies with changing weathers, farming systems and agronomic practices. These variations associated with erratic weather lead to crop loss at vast area and varied extent in different states. Change in climate is likely to increase such risks further, hence use of technology for enrolment, damage assessment and payout of claims is inevitable. Yet, no proper institutional framework is present in insurance sector to develop and monitor these high-tech prospects. A centralized organization is required to synchronize among several stakeholders comprising farmers, insurance agencies, government departments, banks and technology providers. They can be authorized to upkeep, monitor and expand the insurance facilities nationwide.

Some kind of synchronization among the stakeholders also required for use of technology and recommending for further spread. Almost all the state has expertise at agricultural universities/research institutes, regional/state remote sensing centres, ICT players and private agencies. India Meteorology Department has AFMU (Agriculture Field Monitoring Unit) in every agro-ecological region that compiles all climate related events and offers the data to different stakeholders including farmers. A reconnaissance analysis with different stakeholders can offer the opportunity for making an integrated risk management system, which can support them to manage the manifold risks in an articulate manner. This necessitates to establish a robust institutional mechanism at community as well as local government level for supporting the marginal and small farmers to manage risk and make the economy move ahead.

6.5.4 Using Digital Technology: Demand for Image-Based Crop Insurance

For area-based insurance schemes, at least four sites per Gram *Panchayat* (a cluster of villages) need to be selected during harvest to measure the yields for the crop, which is covered under the scheme. This is laborious process to monitor and implement the crop cutting experiments, which generally led to disagreements and delays in settlement of claims. Knowing this, the PMFBY keep arrangements for using advanced technologies through the entire process starting from enrolment to settlement of claims. This comprises using of smartphone images to capture and upload information regarding crop cutting experiments (CCEs); use of satellite images to reduce the number of CCEs that are needed to assess the losses; and the use of drones and remote sensing technologies for providing information on area not sown, mid-season adversities, localized calamities and post-harvest losses. Nevertheless, the operation of remote sensing for yield assessment encounters many challenges, including cloud cover, multi-crops and small size of plots, which necessitates ground truthing of data, which can be provided by ground pictures. In this background, image-based crop insurance acts as a capable digital tool for supporting the activities under existing PMFBY.

6.5.5 Crop Insurance: Improving Business Value Using Technology Interventions

Agriculture sector is so climate-sensitive that hinges exclusively on the probability of weather situations during the crop growing period and that is very improbable due to challenges faced. In an evolving nation like India, where the farming is the primary source of livelihood to marginal and small landholders, their periodic and irregular income laid them in a deprived situation. Calamities like floods, torrential rains and droughts caused by extreme weather lead to huge crop losses throughout the country, which further causes fluctuations in the agri-commodity price. In these adversative circumstances, crop insurance offers financial backing to farmers and makes them eligible for loan during subsequent season. The PMFBY scheme is linking with specialized agencies that uses satellites, remote sensing data, unmanned aerial vehicles (UAVs), artificial intelligence (AI), etc. for assessing crop yield estimates at the panchayat level, removing delays in claim settlements.

On one side, embracing of agri-technological solutions like CropIn's SmartRisk® digitizes many procedures of payouts of claims, starting from enrolment of the farmers on a centralized portal to make sure that they are the right recipients for fair payouts. On the other front, innovative technologies like satellite image processing in association with artificial intelligence and deep learning permit an added accuracy in assessing the farm-lands, curtail the resources requirement throughout the process, and eradicate many limitations of the existing manual processes.

6.6 The Problem of Insurance Demand

Determining the assessment of actual demand for crop insurance products was not attempted and largely remained unsolved, even for developed nations (Vandeveer 2001). While assessing Australian wheat farmers, nearly insignificant willingness to pay over and above the actuarial level of cost, and no buyers was there in many cases, where the loading factor exceeded 20% (Patrick 1988). In another assessment, that too in Australia, farmers were not willing to pay more than 5% of the actuarial cost (Bardsley et al. 1984). Smith and Goodwin (1996, 2010) have evaluated crop insurance in the USA and observed that few farmer's WTP for protection from multi-peril risk was lower in comparison to the costs of providing the insurance, which they stated as reflection of their alternative risk management mechanisms like off-farm employment, diversification or self-insurance rather than risk aversion behaviour. This fact compelled every nation to make any crop insurance scheme as exceedingly subsidized. Even in the developed countries like the USA and Canada, the typical subsidy has remained about 60%, whereas, in Spain and Portugal, subsidies have been near 70%; and in Japan, subsidies have been roughly 50% (Mahul and Stutley 2010; Du et al. 2016). In developing countries, the problem of insurance demand is further perplexed as there are many other mechanisms by which the governments are trying to stabilize farm incomes, such as minimum support price, quotas, input subsidies and low interest agricultural loans (Mahul and Stutley 2010). There are off-course instances of moral hazard, where a blend of high input subsidies, low interest loans and insurance lead to poor management practices in a 'low investment—assured return' setting (Hazell and Hess 2010). Expectation of public food aid reduced the demand for drought insurance, as observed by Sakurai and Reardon (1997) in Burkina Faso. Other risk mitigation measures like grain storage, livestock sales or leveraging social networks are the other factors, which reduces insurance demand in developing nations (Kazianga and Udry 2006; Ambrus et al. 2014). In the situation, where insurance is bundled together with crop loans, low-risk farmers who didn't apply for loan may be hesitant to buy insurance perceiving it cross-subsidizes farmers with higher risks. An unusual situation further obscures the understanding of insurance demand, i.e. loan waiver, which impedes the repayment behaviour and solvency of the banks (Kanz 2016). In Karnataka state, it has been observed recently that loanee farmers hesitate to visit the banks, which are issuing insurance, due to fear of compulsion to refund their credits and they expect that there will be some political intercession during an election time when outstanding loans would be excused (Ghosh 2018). These imply that much focus on estimation of insurance demand may not result equally positive results. There are some instances of estimating insurance demand using non-market valuation methods also, such as the contingent valuation method (CVM) or discrete choice experiments (DCEs). Liesivaara and Myyrä (2017) conducted DCE to include disaster aid as a constant variable to estimate WTP for different characteristics of a crop insurance product in Finland and found that prospects of disaster relief made the farmers less concerned about yield losses. In such cases, premiums need to be heavily subsidized for adopting insurance. Arshad et al. (2016) executed a CVM for eliciting WTP in a theoretical insurance market for two extreme weather events, i.e. floods and droughts and only 28% of respondents were agreeable to choose for insurance, and revealed a very little WTP to the extent of USD 4.49 per year per acre of land for drought and USD 4.72 per year per acre for floods.

6.7 Challenges and Opportunities of Crop Insurance (Subsidies, Incentives, etc.)

First ever crop insurance was introduced by the General Insurance Corporation of India (GICI) in the district of Gujarat in 1972 for H-4 cotton, which was extended to other states like Andhra Pradesh, Karnataka, Maharashtra, Tamil Nadu and West

Bengal. It sustained up to the year 1978 due to major shortcoming of high claim to premium ratio of 8.34. During 1979, GICI in association with State Governments started the Pilot Crop Insurance Scheme (PCIS), on the commendation of Dandekar committee with the coverage against shortfall in yield lower than threshold level. It was implemented in 12 States of India and based on homogenous area approach, which covered cereals, oilseeds, cotton, potato, millets, barley and chickpea (Dandekar 1976). However, it's linked with the crop loans, small and marginal farmers who didn't availed loans were not able to participate and also one major commercial crop like sugarcane was not covered under the scheme. During 1985, Comprehensive Crop Insurance Scheme (CCIS) was started as the first ever scheme spread throughout India, which was also based on homogenous area approach and was linked to short-term credit. The CCIS was initiated with the aim of protecting the farmers against natural catastrophes. Like PCIS, it was also associated to farmers who borrowed loans but CCIS was made compulsory for loanee farmers taking loans for food and oilseed crops.

To expand the insurance coverage in the country, a more inclusive scheme, i.e. NAIS, also known as *Rashtriya Krishi Bima Yojana*, was started in 1999 by replacing CCIS. The scheme was executed by the Agriculture Insurance Company of India Ltd. (AIC) with the aim to cover yield losses from cyclone, flood, hailstorm, drought, pests and diseases, or any other natural disaster. Exceptionality of this scheme was that it pooled both area and individual approach. For localized catastrophes like cyclones, hailstorms, etc., it works on individual approach and for extensive calamities, it works on area approach. The coverage also spread out to all food grains, oilseeds, horticultural and commercial crops for which historical yield data of crop cutting experiments were available and the scheme continued up to *kharif 2013*. The inadequacies of NAIS were investigated, and a MNAIS was started by the Government and major alterations in MNAIS were decrease of unit area to village and panchayat level, changes in premium rates and indemnity level, and modified method of calculations of threshold yield. However, most of the previous crop insurance schemes have remained unsuccessful as they covered multiple perils or provided all-risk coverage (Skees et al. 2001).

Aiming protection of farmers from losses due to weather adversities like excess or deficit rainfall, temperature irregularities, etc., and for bringing more farmers under the coverage of crop insurance, Pilot Weather Based Crop Insurance Scheme (WBCIS) was launched during 2007 to address the issues of risk in coconut. It was considered as potential inheritor of NAIS for covering perennial crop like coconut, which necessitates annual crop practices and is prone to number of biotic and abiotic stresses. PMFBY launched by the government in 2016, which was the most recent yield index and area-based insurance scheme running in India.

Most of the crop insurance schemes in India were based on area approach and the payments were based on estimation of yield loss using CCEs. Therefore, the efficiency of the scheme hinges on the accurateness of the yield assessment from CCE, which is difficult to achieve. So, there is a chance of loss to farmers that does not get compensated which is called as basis risk for the farmer. Further, there is time lag from the institutional end to get compensation owing from so many formalities associated and low coordination among central and state governments (Jain 2004). Low level of education and awareness and complicated jargons in policy terms are other reasons for lack of reception among the farmers (Roberts and Dick 1991). Lack of ground data and varied types of farming practices are also possessing as hurdles in implementation of insurance schemes.

Amid these constraints also, crop insurance sector in the country with full of opportunities as there exists a huge pool of uninsured farmers, who can be brought under coverage with creating awareness and sensitization of their advantages as well as simplification of procedures. This can be achieved by utilizing large network of ICAR institutes including KVKs for dissemination of messages among the farmers. To increase the area and farm coverage multi-agency efforts are required for promoting the crop insurance schemes.

6.8 Gaps in Delivery Mechanism of Major Crop Insurance Schemes and Correcting Imbalances

Most of the crop insurance schemes in India worked on 'area approach', i.e. area is the basic unit for yield loss estimation and computation of compensation to be paid. This works well when there is high correlation among yield during CCEs and farmer's actual yield, and the situation seldom happen that results into 'Basis Risk', i.e. farmers suffered yield loss but do not get compensation. Further, those who finally become eligible to get the compensation, the payment gets delayed substantially in maximum cases. Sharing of loss among central and state governments and procedures involved in this are considered as the main causes for the delay.

As the crop insurance market is not seamless, it is necessary to bundle with other kind of products like credit that has been accomplished in India since the beginning. Whosoever bought short-term crop loan, they routinely and compulsorily insured their crops. However, for a farmer who wish to adopt insurance scheme on voluntary basis, banks need to act as financial mediators. However, the meagre enticement (only 4% of the premium is paid to bank as service charge) that banks received make it unattractive for them in stimulating crop insurance products, which is another reason for less coverage of the schemes. Data indicated that the coverage and indemnity evidently remained biased towards few crops and states. For instance, Gujarat accounted for one-fourth of the total indemnity among the implementing states. Further, greater than one-third of the total indemnity was for a single crop, i.e. groundnut.

Insurance is not an investment to increase income, rather a financial tool for managing the risk, which encompasses difficult jargons like sum insured, indemnity levels, actuarial rates, etc., which need to be clear to farmers for making learned choices. Further, lack of suitable database for fixing premiums and indemnities are also possessing as challenge. A disaggregated study under the NAIS reveals that more than 60% of the beneficiaries, who get compensated for loss from food crops and oilseeds, belong to marginal and small category with landholdings less than two hectares. Further, compensation to these groups accounted for only one-third of the total payouts, due to their small holding size and lower sum assured. In case of horticultural crops, high actuarial rates for annual commercial horticulture crops, non-perennial crops not covered, the inability of the implementing states to consider the village as lower insurance units for minimizing 'basis risk', the disagreeable guaranteed yields and finally, the inability to payout the indemnity on time are the major issues related to the scheme.

The PMFBY and Restructured Weather Based Crop Insurance Scheme are substantially superior to earlier crop insurance schemes, though they have also attracted criticism due to numerous challenges. For instance, the states can now offer add-on coverage for yield loss from wild animals and this added risk alone will require higher premiums to be paid. Further, the deadline for accepting insurance in most of the states is 31st July for *kharif* season. However, by this period, monsoon pattern become clear, particularly in south Indian states. So, in the episode of monsoon shortfall, the state governments are expected to drive for growing insurance coverage to the farmers for getting the claim. In fact, the centre should contend that the state governments decide the move well before the starting of monsoon to avoid the companies be able to factor in the monsoon behaviour.

It is worth mentioning that after so many years of inception, lower than one-fifth of the farmers got insured, with few noteworthy exceptions like Rajasthan, where around 50% of the farmers/holdings are insured. The schemes have not truly evidenced as a significant risk management tool to the farmers in several regions. Instances of farmers' suicides throughout the country implied that measures for managing the risks presently in operation have major deficiencies. Although farmer suicides can be ascribed to a multitude of causes, we cannot refute that an effective insurance would have significantly backed in ameliorating the griefs of the farmers during the period of any catastrophe.

In developing nations like India, the governments whether central or state, cannot compensate for all losses from the risks in agricultural production process. The involvement of government machineries needs to be limited up to augmenting the coping abilities of the farmers, institutional capacities enhancement, support during any catastrophes and fixing imbalances in agricultural sector due to its intrinsic susceptibility. The following initiatives from governments for enhancing risk management capabilities in agricultural sector will lead to ushering in a sustainable and remunerative agriculture:

- Inclusion of technological advancements like electronic weather stations and remote sensing technology, etc. for strengthening the weather insurance system.
- Intensification for dissemination of livestock insurance seeing its potential contribution to rural income growth in future.
- Introduction of farm income insurance schemes to protect farmers' incomes more expansively.
- Introduction of price stabilization fund and credit risks management fund for insulating the farmers from fluctuations in price.

- Managing ecological risks due to variability in climatic events through early warning systems, creating awareness among farmers and institutional mechanism.
- Developing institutions like commodity future markets, contract farming, warehousing infrastructure, etc. to manage the price risks.

6.9 Linkage of Payment to Ecosystem Services (PES) and Crop Insurance

6.9.1 Ecosystem Services: The Concept

Ecosystem services are those services delivered by the environment that benefits the people. As such, there is no solitary, approved process for categorization of all ecosystem services, the Millennium Ecosystem Assessment (MEA) framework is extensively recognized and is considered as a useful beginning. All ecosystem services are providing certain outputs or outcomes, which directly and indirectly influence human well-being, and these thoughts can link well through an economic framework. The assessment of ecosystem services can contribute for superior decision, by taking into consideration the benefits and costs and by emphasizing more clearly the well-being of people. Few of these ecosystem services are well acknowledged including food, fibre and fuel as provisioning services and the cultural services, which offer welfares to people in terms of recreation and aesthetic services. Few other ecosystem services are there, which are not so well known to the people. These comprises the air/water purification, flood protection, climate regulation, soil formation and nutrient cycling.

Ecosystem services contributed to economic welfare through two means—contributing to the income generation and well-being of people and in terms of prevention of losses that impose costs to the society. Through a broader focus on valuing the benefits providing by ecosystems, policies that augment the natural environment are more expected to be considered for investment in creation of natural capital that can make economic sense. Crucial tasks in the assessment of ecosystem services relied on how ecosystems inter-connect for providing services and dealing with high degrees of uncertainty in functioning of ecosystem. This advocates, while assessment is an important tool for designing a policy, it should be considered as the single inputs in decision-making. Methodologies for dealing with the challenges which account for entire ranges of impacts on ecosystems and their services are presently available.

6.9.2 Incentives for Ecosystem Services and Quantification of Its Influence

Ecosystem services boost human well-being through assortment of direct use value (e.g. provisioning services and recreation), indirect use value (e.g. insurance and option value), and non-use value (e.g. existence, intrinsic and bequest value) (Pascual and Perrings 2007). Marketing attributes of the value of ecosystem services rest on the competitiveness of the consumption of the goods/services (whether use by one impedes others to use) and its excludability (whether access is limited to persons paying for it) (Kemkes et al. 2010). Agricultural and livestock produce are competitive and excludable, and are usually valued and merchandized in market places (Farley 2008). Contrarily, public goods (like biodiversity) are non-competitive and non-excludable; common pool resources (e.g. fisheries) are though competitive but non-excludable, and club goods (e.g. entry to eco-park) are non-competitive but excludable (Kemkes et al. 2010). As markets for common pool resources public goods seldom appear automatically, and farmers hardly obtain price for these non-marketed ecosystem services, they remain reluctant to produce it (Ribaudo et al. 2010). Incentives through market-based returns can control the availability of public good and common pool type ecosystem services and correct this market failure (Farley and Costanza 2010). For effectiveness, enticements need to be reinforced by some regulation (Kroeger and Casey 2007), which can be mixed with some incentives and induce the supply of ecosystem services by private individuals (Farley 2008). Different kinds of incentives for ecosystem services are evolving like direct payments/rewards, tax incentives, etc. (Farley and Costanza 2010).

For effective quantification of the influence of incentives on ecosystem services through shift in land use, the relationships need to be understood first. Modelling the influence of incentives on land-use shifting depends on the principle that regional patterns of land use in agroecosystems arise from individual-level land-use choices. Incentives with appropriate regulatory mechanism can alter the relative profitability, which can offer a price signal for shifting the land use by individual landholders (Irwin and Geoghegan 2001; Lewis et al. 2011; Lubowski et al. 2008). For instance, carbon trading can offer commercial prospects for farmers for shifting to tree-based land uses from agricultural land use (Alig et al. 2010; Bryan et al. 2008; Harper et al. 2007). It is indicated that payments for carbon-related trade can induce significant effects in terms of forthcoming benefits through terrestrial carbon sequestration, timber and non-timber forest resources, production of bioenergy, etc. (Alig et al. 2010). Estimation of profitability has been extensively used for evaluating the competitiveness of alternate land uses (Hunt 2008; Maraseni and Cockfield 2011; Wise and Cacho 2011) and quantifying the influence of incentives from ecosystem services (Antle and Stoorvogel 2006; Bryan et al. 2008; Dymond et al. 2012; Polasky et al. 2008; Townsend et al. 2012). However, economic indicators like discount rates, upfront establishment expenses, and current transactions and maintenance costs are also significant factors for judging economic returns from particular land use and their transformation (Bryan et al. 2008).

6.9.3 Payment to Ecosystem Services and Its Working Principle

PES schemes comprise payouts to the landholders in exchange for the supply of a particular ecosystem services over-and-above what would otherwise be paid in the absence of payment system. Payments made by the users of the particular services in context, who may be individual, communities, business entities or government on behalf of other parties. Recipients and land or resource owners may arrive into PES contracts on voluntary basis without any obligation to do it.

Sometimes PES used as an umbrella term to cover whole set of economic arrangements for rewarding the management of ecosystem services. Practically, PES comprises a sequence of payments to landholders or natural resource administrators in return for an assured stream of ecosystem services. Therefore, PES offers a chance to put a value on earlier un-priced ecosystem services like regulation of climate, water quality and the habitat for wild animals and through this, gets them into the larger economic framework. The uniqueness of PES comes from its emphasis on the 'beneficiary pays principle', in contrast to the 'polluter pays principle'. During the last few years a rapid propagation of PES schemes is cropping-up in various countries. Fig. 6.4 presented an illustration of the PES concept in relation to payments for watershed services.

For PES arrangement to work, it is necessary to characterize a win-win for both sellers and the buyers. The PES might be encouraging from a buyer's standpoint, if the expenses are lower than any alternate ways of obtaining the wanted service. For instance, it could be cheaper for a water utility for paying the landowners for better management of catchment compared to paying for added water treatment. The PES

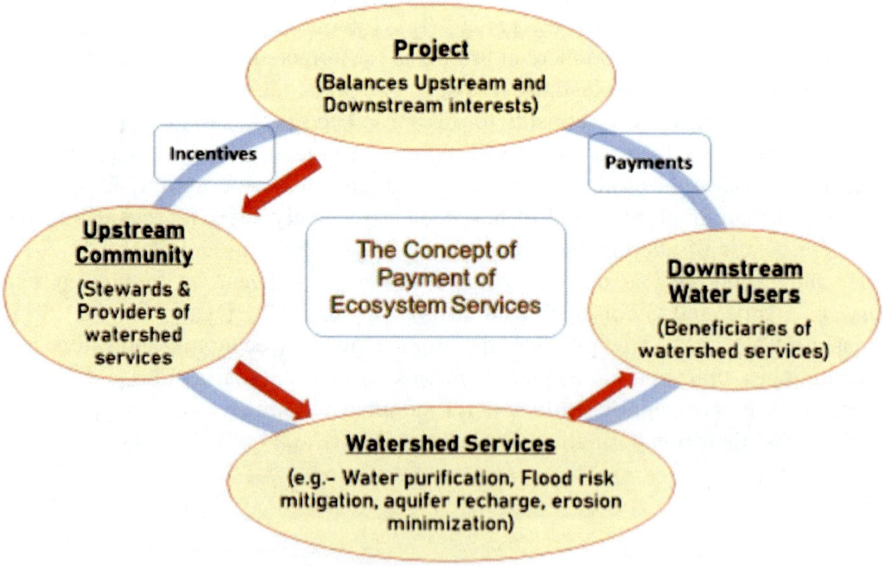

Fig. 6.4 Concept of payment of ecosystem services

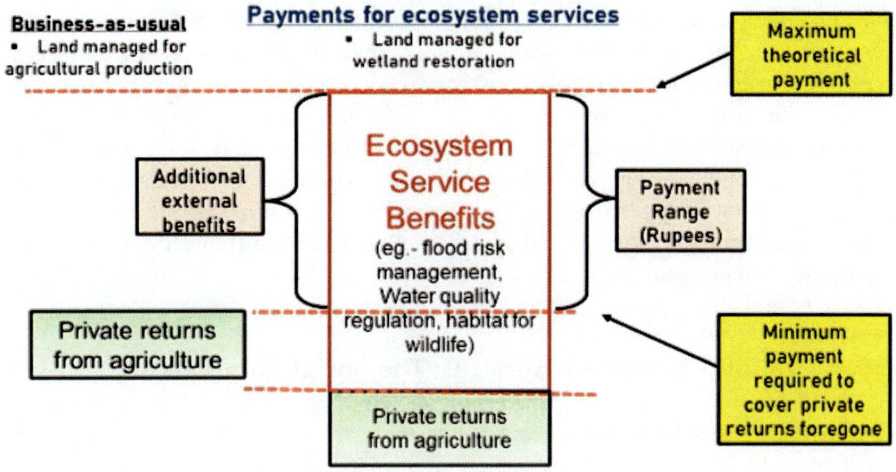

Fig. 6.5 Land managed primarily for agricultural production vs. land managed to provide multiple ecosystem services under PES scheme. Source: Department for Environmental Foods and Rural Affairs 2013

structures may be optimistic from a seller's perception, if the amount of payment obtained at least compensates the worth of any revenues foregone due to implementation of the interventions agreed. For instance, a farmer might be eager to excavate a pond for higher storage capacity, if the anticipated payments at least compensate the digging costs of the pond, including the costs of loss in agricultural production, if any. For illustration, an alteration of an existing cropland to wetland, the minimum PES payment would be usually likely to at least compensate any private return let-off by the farmer/landholder in terms of reduction in agricultural output, the hypothetical highest payment would be the collective worth of added welfares which would be accrued to the buyer(s), and may comprise reduction of flood risk, supply of fresh water, wildlife habitat, etc. (Fig. 6.5). However, many of the benefits are tough to calculate. So practically, the degree at which PES are set would indicate demand and supply for specific ecosystem services and would be agreed-upon at midway between the lowest and highest values. Diverse approaches of land management have differential bearing on delivery of ecosystem service and the allied welfares to individual, communities and businesses establishments. Figure 6.5 also shows the likely processes to augment the delivery of ecosystem services within the farm. A suitably structured PES scheme can deliver the essential inducements for fostering superior land management and improved delivery of ecosystem services.

Alterations in land-use and cropping pattern can also impact the quality of ecological indicators. In comparison to grassland or forest area, agricultural operations normally result greater levels of soil erosion and sedimentation to water body, higher loss of nitrogen to surface as well as groundwater, and lesser amount of soil organic carbon. Similarly, crops also change due to soil erosion, nitrogen loss or soil organic carbon content. Recent research (Claassen et al. 2017) indicated that crop insurance has some influence on quality of environmental parameters in the Corn Belt region (Illinois, Indiana, Iowa, Missouri and Ohio states) and calculated that cropland acreage was 0.06% greater during 2010 with crop insurance in comparison to without it. The impact on crop mix was larger, where crop insurance amplified corn acreage by approximately 1.7%, altering the land away from other crops like wheat, hay and soybeans. These alterations in land use and crop mix triggered about 0.33% surges in soil erosion, nitrogen loss to groundwater (nitrogen loss to surface water was unchanged) and the loss of soil organic carbon content.

6.9.4 Valuing Ecosystem Services: The Social Component

To provide ecosystem services from the farming operations depends on farmers' willingness to adapt the practices, which offer added services and to certain degree that adoption perhaps needs economic compensation in the form of society's willingness to pay for accepting those services. The willingness of farmers to accept new practices, which deliver added services be governed by awareness, attitudes, available resources and also inducements (Swinton et al. 2014). The existing practices are mostly based on the past results; customs; access to technology, policies, and markets, which provide sustained profits. Though environmental stewardship influences choices of many farmers, continued profitability is generally the concern that over-rides the decision.

Earlier sections indicated that providing ecosystem services through agriculture requires inducements. Regardless of the educational level, farmers are conscious about the environmental benefits of alternate practices. Indeed, the farmers who consciously appreciated environmental stewardship were ready to accept less compensation for adopting alternate practices (Ma et al. 2012). However, the larger farmers were keen to receive payments for services, which raises the query *whether consumers are willing to pay for such services?* Whatever may be the mechanism—payments to farmers are inevitable through public or private programmes, abatement of tax, or higher fees to consumers from taxes on polluting inputs or tradable pollution credits (Lipper et al. 2018)—the payments for ecosystem services necessity be borne by the society. Though methods to payment for ecosystem services warrant further research, they are but one among many policy issues existing for meeting the demand for extra ecosystem services.

6.9.5 Payment to Individual Farm at Equilibrium

In economic theory, there is a well-established logical outline for dealing with exterior benefits and costs, in which policy framework is applied to compare marginal 'social benefits' to marginal 'social costs' of any interventions. Marginal social benefits and costs designated the benefits and costs, which are accepted in market, including any outside benefits and costs. This logical framework can be

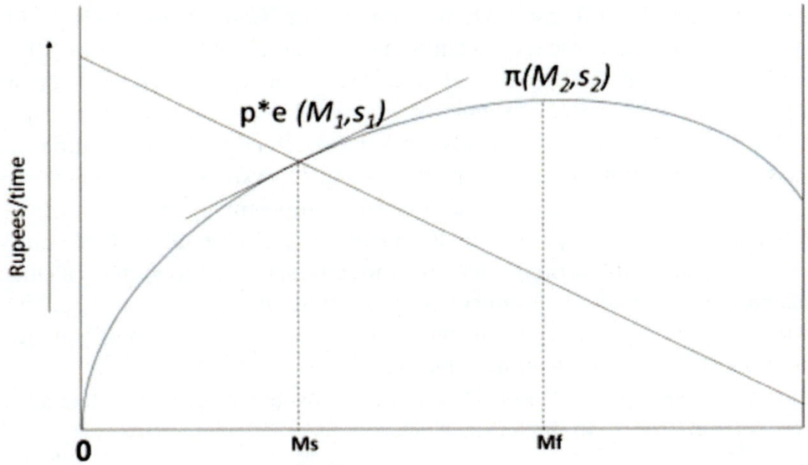

Fig. 6.6 Payment for ecosystem services at the farm level (graphical representation). Note: *M* management activity, *p* price of ecosystem service, *e* ecosystem service, *r* economic returns, M_s socially efficient management, M_f profit maximizing management

elucidated through Fig. 6.6 that shows a diagram estimating the level of a management activity M on the horizontal axis (like, use of a pesticide, or a kind of tillage method, which causes erosion) and economic worth along the vertical axis. The curve r (Ms) denotes farm profits by growing a particular crop at level s, where this curve reaches a highest amount at M_f. The function e (Ms) signifies the magnitude of ecosystem services generated; to make it simpler, it is drawn as a diminishing linear function of M with a highest value reached when $M = 0$. The variable p denotes the price of ecosystem service (rupees per unit of e). The PES to be proficient, p will be equivalent to the marginal social value of the ecosystem service. To appraise the consequences for farmer behaviour, in absence of a payment for the ecosystem services, the farmer would select to produce up to the quantity M_f where profit is highest. But, in case the farmer were paid p rupees per unit of ecosystem services, then the level of farm activity would be equal to [r (M,s) + p e (M,s)]. Consequently, the quantity of M that would be chosen to maximize economic returns to the production activity would be Me < Mf, where the marginal private benefit of M is equated to the marginal benefit of producing more ecosystem service.

6.9.6 Incentives, Land Use, and Ecosystem Services: Synthesizing Complex Linkages

For ecosystem services emanated from agricultural sector, market-based incentives are extensively used for governing the availability and supply, particularly, PES are progressively used to balance the demand and supply of non-marketed amenities. The associations among incentives and land use, and among land use and ecosystem services are complicated. These relationships might be one-to-many or many-to-one, or many-to-many. The associations can be non-linear and differ both spatially and temporally. Based on the incentive design, there can also be feedback on price through changes in the level of supply of ecosystem services. While many studies have addressed individual components, none have actually tried a valuation of the relations among multiple incentives, land uses and ecosystem services in integrated way. So, there is a necessity for sustained development in mainly three areas: quantification of the linkage among incentives and choice of land use change; measuring the effect of shifting land use on ecosystem services, and calculating the incentive price feedbacks from changes in supply of ecosystem service. During recent period, subsidies on crop insurance premiums have been amplified significantly to boost larger participation of farmers.

Sumner and Zulauf (2012) indicated that crop insurance seems to affect agricultural production process in three primary ways:

- The premium subsidies increase the net revenue per unit of land and thereby raise incentives to grow eligible crops and also more of crops with higher subsidy rates.
- The obtainability of crop insurance, which is devised likely by the government programme, encourages cultivating insured crops to more extent that would not otherwise be grown because of the potential risk associated with it.
- Reducing chances of losses due to low yields and/or prices, crop insurance makes incentives for cultivators to accept less of other risk moderating practices.

6.10 Policy-Finance-Technology Amalgamation for Crop Insurance

6.10.1 Policy Issues in Implementing WBCIS

6.10.1.1 Lack of Information

There is lack of information as well as understanding among the clients (farmers) about the sophisticated products like WBCIS, as most of the farmers in India are illiterate. Focus on sensitizing the farmers towards the WBCIS has been overlooked till date and no system has been effectively in use to handle farmers' queries related to the WBCIS. Banks, television and other media options are not used at an expertise level to create awareness and understanding among the farmers for such schemes. So, communicating such schemes to farmers is one of the biggest challenges during the implementation phase.

6.10.1.2 Enrolment Issues

ICT is one sector that keeps developing over time, but yet it has not been utilized properly for ease of enrolment of farmers. Substantial scope exists for using ICT to simplify the enrolment as well as communication to the farmers. Literacy about finance and insurance is important for improving the coverage of the scheme.

6.10.1.3 Affordability

Most of the Indian farmers are small and marginal. Hence, paying high premiums for the insurance may not be possible for them. Another issue is the lack of willingness of farmers to pay premium even when they are able to pay. The subsidy on the premiums by the government solved the later but the former is more important as marginal farmers sometimes are not willing to spend even a small amount of subsidized premiums because of their poor economic condition.

6.10.1.4 Coverage

Unawareness and unsatisfactory experiences of clients with agriculture insurance is an important constraint for wide adaptability of WBCIS to larger part of the community. Non-inclusion of non-loanee farmers, some *rabi* and horticultural crops and livestock under the umbrella of WBCIS is a major issue in extending the coverage of the scheme. Also, the maximum sum insured in this scheme is much less than the cost of cultivation incurred by farmers, which reduces their interest in the scheme and limits its coverage.

6.10.1.5 Product Design

Product available under WBCIS is area specific and applicable to all farmers of that area irrespective of their choices, ability to pay premium and their exposure to the risk. Also, only few weather perils are offered under the scheme, while there are many in the field which can cause damage to crops. Apart from this, variability related to agricultural inputs like water and nutrient availability, varieties, etc. has not been captured under the area approach. All these issues reduce the acceptability of the scheme among farmers.

6.10.1.6 Claim Settlements

Delay in weather data release from the governing authorities like IMD or private institutions sometimes delays the claim settlement of the farmer creating unrest; and also not addressing their queries properly, sometimes creating unpleasant events.

6.10.1.7 Infrastructure

The trust of farmers solely depends on their belief in the credibility of data. Existing infrastructure of AWS and other observatories are not dense enough to provide a very good correlation in the whole RUA. Sometimes, a single AWS serves RUA of radius up to 100 km, which definitely reduces the farmers trust in the data credibility. Such kind of issues needs to be addressed to ensure the trust of farmers in the scheme.

6.10.2 Valuation of Environmental Impacts and Payments for Ecosystem Services in Policy Appraisal

Few of the environmental benefits are easy to value, for example impact of air quality on the level of agricultural output, and the increment in output can be priced using rates in the market. The market price of forest produce may be another instance. However, other instances are also seen, when no direct market for environmental goods and services can be found, which are denoted as non-market goods/services. Nevertheless, there is possibility to attach a price on such environmental goods/services, which can be assessed by various techniques of valuation. Assessing environmental effects allows them to be used for further economic analysis. Such impact can be considered matched to other monetized benefits and costs for establishing the comparative net profit of the proposal/scheme.

Enormous literature reinforces the ecosystem services concept and validates that agriculture interventions do have influences on function of ecosystem and the provisioning of those services. Ecosystem services are natural to the environment that assists all living organisms including human beings through which natural ecosystems and the species that make them up, sustain and fulfil their life (Daily 1997) by cleaning air and water, detoxification and decomposition of waste, restoring fertility of soil, climate regulation, flood and drought mitigation, pests control, pollination of plants, etc. (Salzman et al. 2001). The literature found that land-use pattern by farmers and their management choices may influence physical and biological systems. Some impacts, like alterations in productivity of soil may be restricted to the landholder; others, like runoff from chemical factories into surface waters, may appear off-site. There is abundant indication, which confirmed that due to lack of any inducement strategies, farmers cultivate their plot and make choices based on monetary returns without caring for ecosystem services to the society. For improving the obtainability of ecosystem services outside this private equilibrium, farmers should be provided with inducements for shifting their land-use pattern and management choices, which were creating negative externalities. In most of the cases, ecosystem services are public goods, so some sort of government intervention or transfer of property rights are required. For instance, government can pay the farmers for endorsing the greenhouse gases mitigation by accepting methods that sequester carbon. On the other hand, instruction from government put a limit for greenhouse gas emission for creating a market for 'carbon emissions reduction credits' that could also encourage sequestration of carbon.

6.11 Future Road Map

6.11.1 Opportunities

In spite of having so many hurdles, opportunities exist in the form of huge number of uninsured farmers in the country, who still continuing in farming operations/business. During forthcoming period, lakhs of farmers would be obligated for insuring at least some of their crop area, as the risks in farming activities might intensify in upcoming period due to climate change effects. Diverse climatic settings in the country might also influence to go for insurance at varied extent. The government has already comprehended the distress level with which the farmers are living, and that results into a higher level of policy thrust favouring crop insurance. Lastly, the virtuous linkages of self-help groups and cooperatives prevailing in India can be utilized for micro-insurance, on condition that a decent model is established with appropriate planning and preparation.

6.11.2 Customization of Insurance Products by Region Based on Real-Time Data

Leveraging regional level ground and satellite data for comprehending what grows where and mapping crop regions along with suitable insurance products are the way to customize as per requirement. Knowing the growth and region-wise potential yield can help to modify and modernize prevailing schemes, so that the decisions can be based on real performance data. The following procedures can help to customize the insurance products as the need of the region/people:

- Using information to know the cultivable regions for offering appropriate insurance cover.
- Crop and growth data for the region, which will help for accurate assessment of perils.
- Evaluating risks and opportunities before insuring the crops.
- Analysis of crop health.
- Mapping water stress or water abundance for precise conclusion.

6.11.3 Need for a Nationally Consistent Database

The Government of India has been collecting, collating and publishing data periodically on various aspects of the economy. Government, researchers and others use these databases to understand the past and current situation and also prepare plans for the future. As we move to more an interdependent and globalized world, the quality of databases become an important determinant in reacting to the challenges. As crop insurance is becoming more and more vital in assisting for managing risks at farmer's level. For implementing insurance arrangements, precise data on crop acreage and yield at both individual and collected level are necessary for deciding actuarial premium and weather events and other catastrophes during crop growing period on timely manner are required for deciding the indemnity payments. Presently, the crop insurance is suffering due to absence of both correct information and timely accessibility of the same. Literally 6 months or more crossed for getting CCE data after the harvest of crops. That much delay makes the insurance payment worthless for the farmers who have suffered massive damages and have no other income or support to fall back on. During this present distress situation in agriculture, a decent insurance scheme that can make payment immediately to the farmers can support in alleviating the suffering. As there are no proper system to collect, compile/store and publish the valuable data in time, farmers are continuing to suffer, and insurance as an opportunity to mitigate the distress situations remains unsuccessful. Thus, planning and execution of important programmes remain less effective due to insufficient data generation/publication. Regardless of present scientific developments in capturing, storing and diffusion of data/information, the government does not appear to have paying suitable consideration for using the presently accessible knowledge for evolving a systematic process for data capture, collating, storage and transmission. Hence, adequate efforts are required for making it a national priority for upgrading the database management systems with the aim to support government and other agencies for planning and implementing various policies, programmes and schemes for farmer's benefit.

6.11.4 An Emerging Alternative

- The main cause of failure of the multi-peril crop insurance schemes worldwide is absolute complication of risks and absence of satisfactory risk modelling strategy for understanding these risks. These complications are basically accountable for non-entry of private players in the field of yield insurance in India. Till recently, crop insurance in India was restricted to paying for compensating yield loss. During recent years, this trend has been changed and weather-based insurance products have emerged and popularized. Weather insurance recompenses the indemnities that are not based on the real damages to the insured, instead it based on the weather index, which is extremely associated with actual damages. The index calculated from specific weather variables (like rainfall, wind speed, temperature, relative humidity, etc.)
- The weather insurance scheme was planned after a critical study of the various weather indicators, which affects crop growth during different critical stages—sowing, growth and flowering, grain formation to harvest. Through every stage, the 'trigger' (the level below it weather parameter need to fall for receiving the payments) and 'exit' (the level below which the weather parameter must drop for receiving the maximum payment) are defined. Weather-based insurance has evidently extended the sphere of crop insurance programme in India as insurance now also be available for crops with no past yield data as also for horticultural crops where yield estimates for particular age group are not obtainable. Index-based insurance is not as much of prone to few specific difficulties that are inherent in earlier schemes. The scheme provides benefits to both the insured and the insurer. For the insured farmers, the greatest benefit over the earlier schemes is the scope of getting indemnity payouts timely. The acceptance of weather index schemes also attributed to the clearness as the weather information can be uploaded nearly instantly so that the insured become aware of weather performance in relation to given trigger as well as his eligibility to get the payouts. This scheme also offers a chance to the insured to put in extra efforts or expenses to protect the crop anyway as the compensation would be provided regardless of the yield.
- To escalate the spread, massive awareness drive must be taken up regarding claim structures that presently are more technical and complex. It is even observed that

the insured farmers rarely have any information of the various covers and the degree of weather aberration results to be eligible for claims. The insurers may need to ensure that the message is made known in an articulate way to allow even the uneducated farmers to make informed choices. However, the success of weather insurance would be critically reliant on focusing the challenges mentioned above. An active role of the government is greatly endorsed during initial stages bearing in mind that huge start-up costs involved and the possibility of the insurers shying away from such high investments, even continuing to allow high basis risk. Finally, for the index-based weather insurance to be effective, the insured must be imparted with belief that the index is being calculated precisely and the data is protected from tinkering.

6.12 Messages

Developing nations like India have an enormous market potential for weather index-based crop insurance due to large acreage as well as population involved in crop production. Penetration of index insurance in farming community is very low till now and hence, critically formulated need-based simplified crop insurance products need to be introduced.

There are a number of perils which cannot be protected by any solitary scheme and hence, broad range and multiple schemes should be operated concurrently in the country. In association with the government, participation of private firms should be favoured for providing their services to the farming community. India is so diversified and the needs differ from region to region due to different climate and cropping systems. Therefore, small-scale region-specific products should be developed to cater to the need of local farmers. The area approach is of course superior over the individual approach in many aspects, but the former needs to be scaled down for more acceptability of the data. Linkage among different institutions like banks, government organizations, cooperatives and private firms is necessary to bring down the expenses of the schemes so that profitability of the clients can be improved.

Emerging science of remote sensing and crop modelling, which does crop acreage and yield estimation can be used for improving the weather index insurance. Now with the availability of data with highly spatial, spectral and temporal resolution, it is possible to capture the in-season variability. Moreover, insurance schemes should be framed in such a way to cover all the enterprises of the farm like livestock and fisheries and should be aimed to improve farm income as well as rural economy.

The government must contemplate immediately to establish a committed process within the prevailing institutional set-up or through creation of an innovative independent organization for coordination between different participants covering farmers/groups, industries, various government departments, banks and technology enablers linked to agriculture insurance and associated areas. The said organization should also be authorized for supporting, monitoring and improving insurance facilities countrywide. This might also comprise well-organized data collection, storage and transmission among various stakeholders. Likewise, the state government should contemplate to establish an informal or formal entities for crop insurance, in which specialists from farming, remote sensing, ICTs and insurance business can organize for deliberating the matters connected to crop insurance in the state and recommend the government and other patrons on all connected concerns. Development of advanced insurance schemes with superior technology application in the process of implementations and observing crop insurance would require finance. So, it is suggested that an amount of 0.25% of the premium can be allocated for such drive and should be backed by the central government, state government and insurance agencies.

The new generation agriculture is getting about an uprising in what was once deliberated as a riskiest sector for insurance due to its volatility and outdated methods of cultivation. Digital expertise and precision agriculture are certain recent practices being used for making the cultivation practices more probable, lucrative and sustaining. The capability to trail, monitor and cope with each and every facet of agricultural practices with newer technologies now diminishes the insecurity and provides a protecting shield to farmers in India and the world.

Fast pace of changing climate highlights the instinctive randomness linked to farming. Reasonably, a huge population of the marginal and small group of cultivators have restricted access to worthful crop insurance products. This is the point, where digital farming can create an influence. Backed by AI and Satellite-based information on climate and cultivation details can provide information/forecast about ongoing/upcoming crop sequences, financial organizations can suggest for right kind of investment prospects, and the farmers acquire the access to quality credit and services on timely manner.

Approximately 50 crores small farm families universally represent about 2.5 billion persons who hinges on agrarian output for their sustenance. Yet, 90% among them do not have accessibility to crop insurance that make them prone to massive damages due to drought, flood or other upsetting climatic hazards. Precision farming permits real-time watching of entire farm plots across a region and offers a prism for the impending period. This provides farmers the security they require to counter improbability and permits insurance suppliers to take intended risks while providing credit services.

References

Alig R, Latta G, Adams D, McCarl B (2010) Mitigating greenhouse gases: the importance of land base interactions between forests, agriculture, and residential development in the face of changes in bioenergy and carbon prices. Forest Policy Econ 12:67–75

Ambrus A, Mobius M, Szeidl A (2014) Consumption risk-sharing in social networks. Am Econ Rev 104:149–182

Antle JM, Stoorvogel JJ (2006) Predicting the supply of ecosystem services from agriculture. Am J Agric Econ 88:1174–1180

Arshad M, Amjath-Babu T, Kächele H, Müller K (2016) What drives the willingness to pay for crop insurance against extreme weather events (flood and drought) in Pakistan? A hypothetical market approach. Clim Dev 8:234–244

Aryal J, Sapkota T, Khurana R, Chhetri A, Rahut D, Jat M (2020) Climate change and agriculture in South Asia: adaptation options in smallholder production systems. Env Dev and Sust 20:5045–5075

Babcock BA (2015) Using cumulative prospect theory to explain anomalous crop insurance coverage choice. Am J Agri Econ 97(5):1371–1384

Bardsley P, Abey A, Davenport SV (1984) The economics of insuring crops against drought. Aust J Agric Econ 28:1–14

Barnett BJ, Mahul O (2007) Weather index insurance for agriculture and rural areas in lower-income countries. Am J Agric Econ 89:1241–1247

Bryan BA, Ward J, Hobbs T (2008) An assessment of the economic and environmental potential of biomass production in an agricultural region. Land Use Policy 25:533–549

Claassen R, Langpap C, Wu J (2017) Impacts of federal crop insurance on land use and environmental quality. Am J Agric Econ 99(3):592–613

Clarke DJ, Mahul O, Verma N (2012) Index based crop insurance product design and ratemaking: the case of modified NAIS in India. In: World Bank policy research working paper 5986. Washington D.C, World Bank

Daily GC (1997) Nature's services: societal dependence on natural ecosystems. Island Press, Washington, DC

Dandekar VM (1976) Crop insurance in India. Econ and Pol Weekly 11(26):A61–A80

Department for Environmental Foods & Rural Affairs (2013). Payments for ecosystem services: a best practice guide. Westcountry Rivers Trust. URS 6–8 Greencoat Place, London SW1P 1PL. pp. 84

Du X, Feng H, Hennessy DA (2016) Rationality of choices in subsidized crop insurance markets. Am J Agric Econ 99:732–756

Dymond JR, Ausseil AGE, Ekanayake JC, Kirschbaum MUF (2012) Tradeoffs between soil, water, and carbon–a national scale analysis from New Zealand. J Environ Manag 95:124–131

Elabed G, Carter MR (2015) Compound-risk aversion, ambiguity and the willingness to pay for micro-insurance. J Econ Behav Organ 118:150–166

Fadhliani Z, Luckstead J, Wailes EJ (2019) The impacts of multi-peril crop insurance on Indonesian rice farmers and production. Agric Econ 50(1):15–26

Farley J (2008) The role of prices in conserving critical natural capital. Conserv Biol 22(6): 1399–1408

Farley J, Costanza R (2010) Payments for ecosystem services: from local to global. Ecol Econ 69(11):2060–2068

Feder G, Just RE, Zilberman D (1985) Adoption of agricultural innovations in developing countries: a survey. Econ Dev Cult Chang 33:255–298

Feng H, Du X, Hennessy DA (2020) Depressed demand for crop insurance contracts, and a rationale based on third generation Prospect theory. Agric Econ 51:59–73

Ghosh RK (2018) Performance evaluation of Pradhan Mantri Fasal Bima Yojana (PMFBY). Center for Management in agriculture (CMA), Indian Institute of Management Ahmedabad (IIM-A), Vastrapur, Ahmedabad

Goodwin BK (2001) Problems with market insurance in agriculture. Am J Agric Econ 83:643–649

Harper RJ, Beck AC, Ritson P, Hill MJ, Mitchell CD, Barrett DJ, Smettem KRJ, Mann SS (2007) The potential of greenhouse sinks to underwrite improved land management. Ecol Eng 29:329–341

Hazell PB (1992) The appropriate role of agricultural insurance in developing countries. J Int Dev 4: 567–581

Hazell PB, Hess U (2010) Drought insurance for agricultural development and food security in dryland areas. Food Security 2:395–405

Hazell P, Varangis P (2020) Best practices for subsidizing agricultural insurance. Glob Food Sec 25:100326

Hunt C (2008) Economy and ecology of emerging markets and credits for bio-sequestered carbon on private land in tropical Australia. Ecol Econ 66:309–318

Huo R, Colson G, Ramirez O, Liu Y (2018) Insurance demand over varying coverage levels: experimental evidence in China. Selected Paper, AAEA Annual Meetings, Washington DC, pp 5–7

Irwin EG, Geoghegan J (2001) Theory, data, methods: developing spatially explicit economic models of land use change. Agric Ecosyst Environ 85:7–23. https://doi.org/10.1016/S0167-8809(01)00200-6

Jain RCA (2004) Challenges in implementing agriculture insurance and re-insurance in developing countries. The Journal:14–23

Jensen ND, Mude AG, Barrett CB (2018) How basis risk and spatiotemporal adverse selection influence demand for index insurance: evidence from northern Kenya. Food Policy 74:172–198

Kanz M (2016) What does debt relief do for development? Evidence from India's bailout for rural households. Am Econ J Appl Econ 8:66–99

Kazianga H, Udry C (2006) Consumption smoothing? Livestock, insurance and drought in rural Burkina Faso. J Dev Econ 79:413–446

Kemkes RJ, Farley J, Koliba CJ (2010) Determining when payments are an effective policy approach to ecosystem service provision. Ecol Econ 69:2069–2074

Kiran S, Umesh KB (2015) Willingness to pay for crop insurance premium: a study on maize farmers in India. In: Conference of International Association of Agricultural Economists, Milan, Italy, August 9–14, 2015 (No. 1008–2016-80222), pp 1–13

Kroeger T, Casey F (2007) An assessment of market-based approaches to providing ecosystem services on agricultural lands. Ecol Econ 64:321–332

Lewis DJ, Plantinga AJ, Nelson E, Polasky S (2011) The efficiency of voluntary incentives policies for preventing biodiversity loss. Resour Energy Econ 33(1):192–211

Liesivaara P, Myyrä S (2017) The demand for public–private crop insurance and government disaster relief. J Policy Model 39:19–34

Lipper L, McCarthy N, Zilberman D, Asfaw S, Branca G (2018) Climate smart agriculture: building resilience to climate change. Springer International Publishing, Cham, Switzerland

Lubowski RN, Plantinga AJ, Stavins RN (2008) What drives land-use change in the United States? A national analysis of landowner decisions. Land Econ 84:529–550

Ma S, Swinton SM, Lupi F, Jolejole-Foreman MCB (2012) Farmers' willingness to participate in payment-for-environmental-services programmes. J Agric Econ 63:604–626

Mahul O, Stutley CJ (2010) Government support to agricultural insurance challenges and options for developing countries. The World Bank, Washington DC, p 2010

Maraseni TN, Cockfield G (2011) Crops, cows or timber? Including carbon values in land use choices. Agric Ecosyst Environ 140:280–288

Miranda MJ, Farrin K (2012) Index insurance for developing countries. Appl Econ Perspect Policy 34:391–427

Morduch J (2006) Micro-insurance: The next revolution? In: Banerjee A, Benabou R, Mookherjee D (eds) Understanding poverty. Oxford University Press, Oxford, pp 337–356

Pascual U, Perrings C (2007) Developing incentives and economic mechanisms for in situ biodiversity conservation in agricultural landscapes. Agric Ecosyst Environ 121(3):256–268. https://doi.org/10.1016/j.agee.2006.12.025

Patrick GF (1988) Mallee wheat farmers' demand for crop and rainfall insurance. Aust J Agric Econ 32:37–49

Polasky S, Nelson E, Camm J, Csuti B, Fackler P, Lonsdorf E, Montgomery C, White D, Arthur J, Garber-Yonts B, Haight R, Kagan J, Starfield A, Tobalske C (2008) Where to put things? Spatial land management to sustain biodiversity and economic returns. Biol Conserv 141:1505–1524

Ribaudo M, Greene C, Hansen L, Hellerstein D (2010) Ecosystem services from agriculture: steps for expanding markets. Ecol Econ 69:2085–2092

Roberts R, Dick WSA (1991) Strategies for crop insurance planning, FAO agricultural services bulletin 86. FAO, Rome

Sakurai T, Reardon T (1997) Potential demand for drought insurance in Burkina Faso and its determinants. Am J Agric Econ 79:1193–1207

Salzman J, Thompson BH, Daily GC (2001) Protecting ecosystem services: science, economics, and policy. Stanford Environmental Law Journal 20:309–332

Skees JR (2000) A role for capital markets in natural disasters: a piece of the food security puzzle. Food Policy 25:365–378

Skees J, Gober S, Varangis P, Lester R, Kalavakonda V (2001) Developing rainfall based index insurance in Morocco. Policy research working paper 2577. World Bank, Washington DC

Smith VH, Glauber JW (2012) Agricultural insurance in developed countries: where have we been and where are we going? Appl Econ Persp and Policy 34(3):363–390

Smith VH, Goodwin BK (1996) Crop insurance, moral hazard, and agricultural chemical use. Am J Agric Econ 78:428–438

Smith VH, Goodwin BK (2010) Private and public roles in providing agricultural insurance in the United States. In: Brown JR (ed) Public Insurance and Private Markets. AEI Press, Washington, DC, pp 173–210

Smith VH, Watts M (2010) The new standing disaster program: a SURE invitation to moral hazard behavior. Appl Econ Perspect Policy 32:154–169

Sumner DA, Zulauf C (2012) Economic & environmental effects of agricultural insurance programs. The Council for Food, Agriculture and Resource Economics, Washington DC

Swinton SM, Rector N, Robertson GP, Jolejole-Foreman MCB (2014) Farmer decisions about adopting environmentally beneficial practices. In: Hamilton SK, Doll JE, Robertson GP, Lupi F (eds) The ecology of agricultural ecosystems: long-term research on the path to sustainability. Oxford University Press, New York, NY

Townsend PV, Harper RJ, Brennan PD, Dean C, Wu S, Smettem KRJ, Cook SE (2012) Multiple environmental services as an opportunity for watershed restoration. Forest Policy Econ 17:45–58

Vandeveer ML (2001) Demand for area crop insurance among litchi producers in northern Vietnam. Agric Econ 26:173–184

Wise RM, Cacho OJ (2011) A bioeconomic analysis of the potential of Indonesian agroforests as carbon sinks. Environ Sci Pol 14:451–461

Wu S, Goodwin BK, Coble K (2019) Moral hazard and subsidized crop insurance. Agric Econ 51: 131–142

Yu J, Sumner DA (2018) Effects of subsidized crop insurance on crop choices. Agric Econ 49(4): 533–545

Glossary

Bioremediation Pricing Bioremediation is the green technology that involves organic substance, microbes and tolerant crops (phytoremediation) to clean up the groundwater, air or soil from toxic pollutants. Pricing of bioremediation could be done in terms of pollution control reducing hazards and providing ecosystem services.

Carbon exchanges These are the balance sheets of carbon inflows or sequestration and carbon emission in a system.

Carbon sequestration Carbon sequestration in agriculture refers to the storing of carbon for relatively longer period (10–10,000 years) in plant and soil systems.

Challenges of pricing of ESs in agriculture Defining the ownership of services and subsequent enforcement; and the absence of market for most of ESs, Uncertainty on environmental performance; uncertainty reliable off-sets; long-term nature of the benefits; and lacking precise methodologies are the other associated challenges.

Cooperative approach for valuation of ES in agriculture In this approach, a collective management based on the resources characterization for providing optimum ES is done. It has a logical framework of interrelated activities for a particular ES or the flows of ES at different scale having long-term goal and management-focus on the non-production-related services.

Cost It is the price of something that we would be expected to pay for some goods or services.

Value It is the quality of an object (goods/services; both tangible and intangible) that permits measurability and therefore comparability. Valuation in an economic context can be particularly helpful for comparing systems with a complex set of socio-ecological relationships, often the case with ecosystems.

Cultural services The ecosystem services which enhance the quality of life of human being are termed as cultural services. Those services primarily include aesthetic, recreation, cultural identity, etc.

Double counting of ecosystem services It refers to a wrong practice of counting the value of an ecosystem service more than one time. It generally happens when spatiotemporal scale of beneficiary is not identified properly and misinterpretation of final end-product of particular service.

P. Bhattacharyya et al., *Pricing of Ecosystem Services in Agriculture: A Basis of Crop Insurance*, https://doi.org/10.1007/978-981-19-4416-1

Economic and environmental losses of straw burning The losses include nutrients losses, yield losses to the next crop, additional irrigation charges, emissions of GHGs and toxic pollutants, biodiversity losses and human health hazards.

Ecosystem services provided by agriculture The ecosystem services provided by agriculture are provisional (food, fodder, fibre, fuel, raw materials for industries, by-products), regulatory (gas and water regulation, erosion control, pollination, flood control, etc.), supporting (soil fertility, soil formation, nutrient cycling, hydrological flow, etc.) and cultural (aesthetic, recreation, etc).

Ecosystem services The benefits that human beings get from the effective functioning of the ecosystem. Ecosystem functioning and/or processes directly or indirectly contribute to human well-being.

Ecosystems disservice in agriculture The services that generally increase the production cost and/or reduce productivity, for example, weed competition with desired crops for nutrients and water, losses of beneficial insects can cause outbreaks of certain pests and diseases and nutrients losses through runoff.

Eddy Covariance technique A micro-meteorological technique which primarily measures the wind velocity in three directions (by sonic anemometer) and scalars (gas concentration, CO_2, methane, etc.) and estimates the fluxes by calculating the covariance of the wind velocity at Z direction and the scalar.

Gas regulation in agriculture Gas regulation in agriculture takes place either by emitting oxygen and absorbing CO_2 (as sink) during photosynthesis-respiration process or by emitting GHGs (non-CO_2 greenhouse gases like CH_4 and NO_2) as source during soil-respiration and other metabolic biochemical processes in soil-microbe-plant-atmosphere systems.

General method of pricing of ESs in agriculture The common method follows for assessment of ESs in agriculture is summing up the values of all individual ES including provisioning, regulatory, supporting and cultural services.

Greenhouse gases The GHGs are the gases in earth's atmosphere which can absorb and radiate back the longwave radiations and keep our planet warm. The major GHGs related to agriculture are carbon dioxide, methane and nitrous oxide.

Importance of pricing of ecosystem services The pricing/valuation of ecosystem services is to promote awareness among all the stake holders (farmers, policy makers, civil citizens, industries, government staff and officials) regarding the concept and importance of ESs and mainstreaming those within the government policy framework.

Indirect values Those are associated with the services provided by nature that are not directly consumed, often being associated with regulating ecosystem services.

Natural capital It is derived from those ecosystem services which are not built or maintained by human activities, i.e. natural ecosystem and its products, but its actual benefit could be obtained when combined with 'built', 'human' and 'social' capital.

Offsets of ESs in agriculture Measurable service/conservation outcomes generated due to certain ecosystem services (non-tangible) provided by agriculture. Those are also used to quantify the adverse impacts on ecosystem if services are withdrawn or not provided. The 'offsets' should be real and cheaper than the mitigation costs fixed by regulated authorities.

Option values Those are the values people place on having the option to enjoy something in the future even if they do not currently use it; this can be particularly important in the case of land, soil and water, which passed down through the generations.

Peatlands and ecosystem services Peatlands provide ecosystem services in terms of carbon sequestration, biodiversity maintenance, water recharging and recreational benefits. Valuation of ESs provided by peatland can be indirectly done through restoration cost of degraded peatland with specific time frame.

Price It is the amount of money we pay for certain marketable goods and services.

Pricing of biological control of pests These ecosystem services are quantified indirectly by estimating the 'predation rate', through quantifying the predators and parasitoids for a particular agricultural system.

Pricing of carbon sequestration in agroforestry The carbon sequestration price of agroforestry system can be calculated by 'partial market equilibrium' considering the 'average cost-curve' and 'economic break analyses' to fix the 'supply cost'.

Pricing of ecosystem services Valuation of ecosystem services in monetary term. Putting the monetary values of ESs in economic term which can be marketable.

Pricing of nitrogen fixation in agriculture The total amount of nitrogen fixed per unit land (either by beneficial nitrogen fixing microorganism or by leguminous crops) multiplied with the actual amount of nitrogenous fertilizer that provides the amount of nitrogen and the unit price of the fertilizer.

Pricing of plant nutrient mineralized The pricing of plant nutrients mineralization in soil can be calculated by converting the total amount of nutrients mineralized to their equivalent price of fertilizer cost.

Pricing of soil fertility The pricing of soil nutrient supply can be done by multiplying the nutrient uptake by plant from soil and to the unit price of respective fertilizers contributing to the specific nutrients.

Pricing of soil formation The pricing of soil formation can be done by multiplying the value of topsoil with the total quantity of soil formed annually.

Provisioning services by agriculture The services provided by agriculture through the production of food, fodder, fibre, fuel, by-products and raw materials for industries which have economic values and can be directly traded in markets.

Provisioning services Ecosystem services that provide food, fodder, timber and raw material for any production system and industries or 'other provisioning'-- benefits which are usually made up of 'human', 'built' and 'social' capitals.

Purpose of pricing of ESs in agriculture Pricing of ESs in agriculture would promote agricultural development along with ecosystem sustainability. Pricing of ESs in agriculture would sensitize policy makers and the other stake holders regarding the values accrued to the society as a whole and needs for maintaining those.

Rainwater harvesting ponds (RHP) and ecosystem services The RHPs are the anthropogenic wetlands that provide ESs in terms of carbon sequestration, regulation of hydrology, maintenance of biodiversities and cultural services.

Regulating services Ecosystem services which regulate different aspects of natural and human resources/processes on integrated systems. For example, water regulation and purification, flood control, human disease prevention and protection, cyclone-protection and air quality-maintenance.

Rice straw management and ecosystem services The environmental gains of different alternative utilization of straw (bioethanol production, biochar conversion, mushroom production, compost preparation, etc) can be considered as ecosystem services provided by the certain residue management practices.

Seagrass and ecosystem services The ESs provided by seagrass systems are coastal bank protection, coastal erosion control, maintaining diversities of marine life and fisheries and 'blue carbon' sequestration.

Soil ecosystem services Soils are an important natural capital that has specific functions providing multiple ESs, which depend on their properties and influenced by its use and management. Soils help in filtering and cleaning of drinking water; regulate water flow; deliver plant nutrients and host microorganism that decompose organic matter.

Soil natural capital It is defined by the capacity of soil to provide the ecosystem services required for a determined land use, assuming that sustainable practices are being used.

Soil organic carbon inputs in agriculture The SOC inputs include litter deposition, manure application, roots, stubble residues left over in field after harvest and rhizodeposition including root exudates.

Supporting services Ecosystem services that support and maintain the basic ecosystem functions and interlinked processes like nitrogen and carbon fixation, formation of soil and maintenance of soil health, sustaining the habitat of flora and fauna.

Total economic value (TEV) It is the sum of all relevant use and non-use values generated now and, in the future, that is, the sum of the producer and consumer surplus under the demand curve, excluding the cost of production.

Valuation It is the process of expressing one of the qualities of an action or object on a scale.

Carbon footprint of energy generation The carbon footprint of any energy generation system or device can be calculated from the emission factor (EF) of the method of energy generation in the system after dividing it with their efficiency.

Internal rate of return (IRR) The IRR is an economic metric used to calculate the profitability of any investment. The IRR is defined as the discount rate at which Net Present Value (NPV) equals to zero at particular time within its life cycle period.

Solar thermal technology (Principle) The principle of solar thermal technology is the generation of heat by trapping the incident solar irradiation inside an insulated enclosure chamber through a transparent glass collector and further using the generated heat energy for different purposes, for example, drying, heating and cooking.